International Scientific Seminars on Fundamental and Applied Problems of Photonics and Condensed Matter Physics 2014

AA009052

Journal of Physics: Conference Series Volume 584

Moscow, Russia
30 May – 27 June 2014

ISBN: 978-1-63439-915-9
ISSN: 1742-6588

Printed from e-media with permission by:

Curran Associates, Inc.
57 Morehouse Lane
Red Hook, NY 12571

Some format issues inherent in the e-media version may also appear in this print version.

Copyright© by the Institute of Physics
All rights reserved.

Printed by Curran Associates, Inc. (2015)

For permission requests, please contact the Institute of Physics
at the address below.

Institute of Physics
Dirac House, Temple Back
Bristol BS1 6BE UK

Phone: 44 1 17 929 7481
Fax: 44 1 17 920 0979

techtracking@iop.org

Additional copies of this publication are available from:

Curran Associates, Inc.
57 Morehouse Lane
Red Hook, NY 12571 USA
Phone: 845-758-0400
Fax: 845-758-2634
Email: curran@proceedings.com
Web: www.proceedings.com

TABLE OF CONTENTS

Journal of Physics: Conference Series **584** (2015) 012001 doi:10.1088/1742-6596/584/1/012001

International Scientific Seminars "Fundamental and Applied Problems of Photonics and Condensed Matter Physics", Bauman Moscow State Technical University, Moscow, Russia, May – June , 2014

International Scientific Seminars "Fundamental and Applied Problems of Photonics and Condensed Matter Physics" were held in Bauman Moscow State Technical University (BMSTU) in May - June 2014. The idea of the Seminars was to organize a series of meetings between young scientists and discuss actual problems and the latest results in Photonics and Condensed Matter Physics. There were eight Sessions:

Modern Problems of Condensed Matter Physics;
Laser Physics;
Spectroscopy of Condensed Matter;
Terahertz Optical Technology;
Optical Signals Processing;
Physics of Optical Strong Correlated Systems;
Complex Dusty Plasma Physics;
Biomediacal Applications of Photonics.

Seminars were organized by the young group of scientists and students from Research and Educational Center "Photonics and Infrared Technology" at BMSTU. It brought a significant contribution to the development of youth science in the field of Physics and Photonics in Russia.

More than 100 young scientists and students participated in the Seminars in spring - summer 2014.

The International Scientific Seminars were supported by the Russian Foundation for Basic Research (grant # 14-08-06030-g).

This volume contains proceedings of the International Scientific Seminars "Fundamental and Applied Problems of Photonics and Condensed Matter Physics".

Stanislav Yurchenko and Viktor Ryzhii

Bauman Moscow State Technical University

Content from this work may be used under the terms of the Creative Commons Attribution 3.0 licence. Any further distribution of this work must maintain attribution to the author(s) and the title of the work, journal citation and DOI.
Published under licence by IOP Publishing Ltd

Journal of Physics: Conference Series **584** (2015) 012001 doi:10.1088/1742-6596/584/1/012001

Seminar Chair
Stanislav O. Yurchenko, Leading Researcher at Bauman Moscow State Technical University (Russia), PhD in Physics

Adviser
Victor I. Ryzhii, Chief Researcher of the Research and Educational Center "PHOTONICS AND INFRARED TECHNOLOGY", Prof. at Tohoku University (Japan)

РФФИ
РОССИЙСКИЙ
ФОНД
ФУНДАМЕНТАЛЬНЫХ
ИССЛЕДОВАНИЙ

МГТУ
1830

Peer review statement

All papers published in this volume of *Journal of Physics: Conference Series* have been peer reviewed through processes administered by the proceedings Editors. Reviews were conducted by expert referees to the professional and scientific standards expected of a proceedings journal published by IOP Publishing.

Content from this work may be used under the terms of the Creative Commons Attribution 3.0 licence. Any further distribution of this work must maintain attribution to the author(s) and the title of the work, journal citation and DOI.
Published under licence by IOP Publishing Ltd

*** Formatting Issue - Best Available Paper/Graphic ***

BMSTU Seminars IOP Publishing
Journal of Physics: Conference Series **584** (2015) 012001 doi:10.1088/1742-6596/584/1/012001

Automated screening of pigmentary skin neoplasms

Konstantin G Kudrin[1*], **Oleg V Matorin**[2] **and Igor V Reshetov**[2]

[1] Institute of Improvement of Professional Skill of the Federal Medico-Biological Agency of Russia, 125371, Volokolamskoye shosse 91, Moscow, Russia

[2] P. A. Hertzen Moscow Research Institute of Oncology, 125284, 2nd Botkinsky proezd 3, Moscow, Russia

E-mail: *kudrin_k@rambler.ru

Abstract. We have analysed the clinical symptoms and the malignization signs of pigmented skin neoplasms. We have estimated the complex of clinical parameters which could be measured for the purpose of skin screening diagnostic via digital image processing. Allowable errors of clinical parameter characterization have been calculated, and the origin of these errors has been discussed. Proposed technique for automated screening of pigmentary skin neoplasms should become an effective tool for early skin diagnostics.

1. Introduction

Skin neoplasms represent 39% of total derma-cosmetic morbidity, and melanoma is the most dangerous skin disease. The incidence of melanoma is not large comparing to the incidence of other skin tumors. It varies at different regions of the world, averagely amounting 3-5% of all skin tumor incidence. This neoplasm is the one, which most frequently causes the death of the patients suffering from skin cancer [1,2]. Thus, the problem of pigmentary skin neoplasm (PSN) screening and early melanoma diagnosis is of high importance in oncology, modern medical sciences, biomedical technologies, and biomedical optics.

Standardized incidence rates of melanoma increase annually by 7% over the last 20 years in most countries of the world [3,4]. Every year approximately 7000 melanomas are being identified in Russia. In 83% of all cases the tumor already reaches high levels of invasion in underlying tissue by the beginning of treatment. Unfortunately, late diagnosis of melanoma significantly affects the performance of the five-year survival rate [5]. One of the effective methods of early melanoma detection is screening of PSN [6]. PSN screening involves both the identification of patients, having suspicious neoplasms with high risk of melanoma development, and the identification of such neoplasms in a patient, having the syndrome of skin polyneoplasia. Besides the adequate diagnostic efficiency, methods and instruments for PSN screening should have enough speed, defined as the surface skin area being examined at the appropriate time interval. Therefore, studying skin by means of digital imaging, combined with appropriate image processing techniques, seems to be fast and effective approach for skin screening diagnosis. Such systems allow registration of so-called body maps, skin maps, and mole maps, and they could be classified as automated systems for skin neoplasms screening (ASSNS).

In the present paper we are examining the theoretical background for PSN screening by means of ASSNS. We are considering a list of problems, including: numeric description of PSN, allowable errors for determination of PSN clinical parameters, and sources of PSN determination errors.

Content from this work may be used under the terms of the Creative Commons Attribution 3.0 licence. Any further distribution of this work must maintain attribution to the author(s) and the title of the work, journal citation and DOI.
Published under licence by IOP Publishing Ltd

*** Formatting Issue - Best Available Paper/Graphic ***

2. Clinical parameters of pigmented neoplasms

The skin melanoma is the most dangerous skin tumor [7], and the problem of PSN screening and early melanoma diagnosis remains unresolved despite the wide range of diagnostic tools [8, 9], including spectroscopy [10, 11, 12, 13], dermatoscopy [14, 15, 16, 17], ultrasonography [18, 19], confocal scanning laser microscopy [20, 21], OCT imaging [22]. Thus, development of novel approaches for effective melanoma diagnosis is of high importance [23] and this confirms the search for new melanoma diagnostics methods [24]. Modern global trend in early melanoma diagnosis is associated with the use of ASSNS systems for this purpose [25].

Principles of ASSNS work are based on the methods of digital image processing and morphometrics. Combining the appropriate hardware and software in a diagnostic systems we are able to successfully solve the problem of numeric estimation of clinical signs, corresponding to the different skin lesions. Modern diagnostic systems implement the skin neoplasms screening by means of mapping the patient skin surface measure only PSN sizes and surface area, shape factor and relative brightness [26]. Obviously, this list of parameters is clearly insufficient to estimation of the malignization signs and PSN clinical signs, well-known and widely-used in medical practice.

Clinical description of pigmented skin neoplasms implicates information about sizes, shape, colour, border features, and surface texture [27, 28]. All techniques for early melanoma detection, which are widely utilized in clinical practice [29], for example, the ABCD rule [30], allow expertly and quantitatively (in balls) estimate the size (diameter), shape (asymmetry), color (color inhomogenioty), and border features (sharpness, irregularity). Moreover, the following list of visual malignization signs of pigmented skin neoplasms is well known in dermatology [31]:

- missing skin pattern on the surface;
- appearance of shiny, glossy surface;
- appearance of asymmetry or abnormal shape; shape change;
- appearance of rugged border;
- horizontal growth;
- surface peeling with dry crusts formation;
- absence or loss of hair on surface;
- partial (inhomogeneous) or a complete change of color; appearance of depigmentation area;
- appearance of small nodules on the surface;
- vertical growth over the surrounding tissues.

The visual signs and malignization signs are grouped in categories, describing:

- sizes;
- form;
- border;
- color and pigmentation homogeneity;
- changes in size, shape, border, color and pigmentation homogeneity.

Figure 1 shows the clinical parameters describing PSN. These PSN features could be automatically determined utilizing the methods of biomedical image morphometry.

*** Formatting Issue - Best Available Paper/Graphic ***

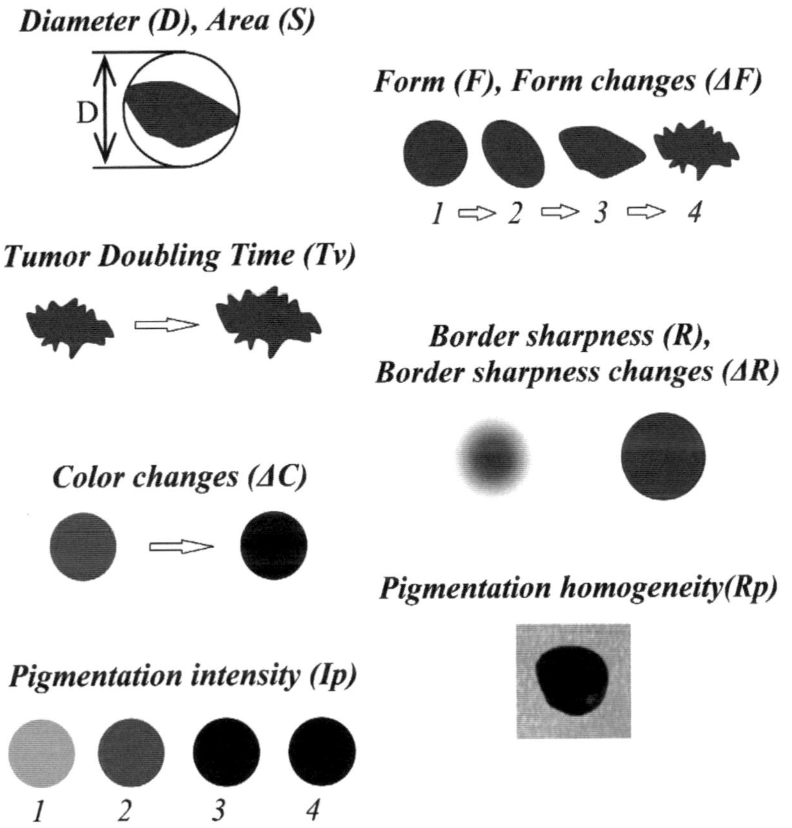

Figure 1. PSN clinical parameters for automatically measuring by means of biomedical image morphometry.

3. Estimating the allowable errors of PSN clinical parameter measurement

Except the sizes of PSN (its diameter and area), all clinical parameters exhibit qualitative character (figure 1). Therefore, for automatic estimation of clinical parameters we need to measure a complex of PSN features, including:

- maximal diameter;
- PSN areas corresponding to the current and previous examination;
- period of time between examinations;
- ratio of maximal diameter to maximal transverse distance between the border points;
- relative deviation of PSN area from the area of equivalent ellipse;
- relative deviation of PSN border length from the length of smooth border;
- normalized gradient of border intensity;
- contrast;
- standard deviation of color coordinates;
- average value of color coordinates for current and previous examination.

*** Formatting Issue - Best Available Paper/Graphic ***

Table 1. The relationship of the complex clinical and measured parameters.

Parameter	Expression	Measured parameters	
		Symbol	Name
D	$D = L$	L	maximal diameter
S	$S = S_{н}$	$S_{н}$	area for current examination
T_S	$T_S = \dfrac{T}{\dfrac{S_{н}}{S_{с}} - 1}$	$S_{с}$	area for previous examination
-	-	T	time between examinations
F	$F = \begin{cases} 1, E < P_E, \delta S_E < P_{\delta S_E}, \delta L < P_{\delta L} \\ 2, E \geq P_E, \delta S_E < P_{\delta S_E}, \delta L < P_{\delta L} \\ 3, \delta S_E \geq P_{\delta S_E}, \delta L < P_{\delta L} \\ 4, \delta L \geq P_{\delta L} \end{cases}$	E	ratio of maximal diameter to maximal transverse distance between the border points
-	-	δS_E	relative deviation of PSN area from the area of equivalent ellipse
-	-	δL	relative deviation of PSN border length from the length of smooth border
ΔF	$\Delta F = \begin{cases} 1, F = F_c \\ 2, F \neq F_c \end{cases}$ F_c – Form for the previous examination	–	–
R_g	$R_g = \begin{cases} 1, Grad \geq P_{Grad} \\ 2, Grad < P_{Grad} \end{cases}$	$Grad$	normalized gradient of border intensity
ΔR_g	$\Delta R_g = \begin{cases} 1, R_g = R_{gc} \\ 2, R_g \neq R_{gc} \end{cases}$	–	–

*** Formatting Issue - Best Available Paper/Graphic ***

	R_{g_c} – Border sharpness for the previous examination		
I_p	$I_p = \begin{cases} 1: K < P_{1K};\ 2: P_{1K} \le K < P_{2K} \\ 3: P_{2K} \le K < P_{3K};\ 4: K \ge P_{3K} \end{cases}$	K	contrast
R_p	$sI = 100\ \% \cdot \frac{1}{3}\left(\frac{sr}{r_н} + \frac{sg}{g_н} + \frac{sb}{b_н}\right)$ $R_p = \begin{cases} 1, sI < P_{sI} \\ 2, sI \ge P_{sI} \end{cases}$	sr, sg, sb	standard deviation of color coordinates
ΔC	$\delta C = 100\ \% \cdot \frac{1}{3}\left(\frac{\lvert r_н - r_c\rvert}{r_r} + \frac{\lvert g_н - g_c\rvert}{g_c} + \frac{\lvert b_н - b_c\rvert}{b_c}\right)$ $\Delta C = \begin{cases} 1, \delta C < P_{\delta C} \\ 2, \delta C > P_{\delta C} \end{cases}$	$r_н, g_н, b_н$ r_c, g_c, b_c	average value of color coordinates for current and previous examination

Table 2 shows the relationship between measured parameters and complex clinical parameters of PSN.

Table 2. Allowable errors of measured parameters.

Measured parameters	Allowable errors	
Symbol	Equation	Value
x	$\Delta_L = \frac{L \cdot \Delta T_S}{2 \cdot T_S}\sqrt{\frac{\lvert T^2 - T_S^2\rvert}{T^2 + T_S^2}}$	0,45mm
$S_н$ S_c	$\Delta_S = \frac{S_t - S_{t-T}}{T}\sqrt{\frac{\Delta_{T_t}{}^2 (S_t - S_{t-T})^2 - \Delta_T{}^2 S_{t-T}{}^2}{S_{t-T}{}^2 + S_t{}^2}}$	2,9 mm^2
T	Δ_T	0,5 days
E	$\Delta_E = \frac{\Delta_L}{L} \cdot \sqrt{2}$	0,30
δS_E	$\Delta_{\delta S} = \frac{\Delta T_S}{T}\delta S\sqrt{\lvert \delta S^2 - 1\rvert}$	0,117

*** Formatting Issue - Best Available Paper/Graphic ***

δL	$\Delta_{\delta L} = \frac{\Delta_L}{L} \cdot \sqrt{2}$	0,30
$Grad$	$\Delta_G = \frac{\Delta_L}{\Delta X^2}$	7,5%
K		
sr, sg, sb $r_н, g_н, b_н$ r_c, g_c, b_c	Research results	4%

Based on existed published data on the parameters of PSN size and the PSN changes, significantly important for diagnostics [28, 32], we have obtained allowable errors for estimating linear and area measured parameters (table 1), utilizing the methodology of error calculation for the indirect measurement case. Results of allowable error calculation are presented in the table 2. In order to estimation the allowable errors for measuring the color parameters (table 1) we use modern data on the capabilities of the human visual analyser [33]. Table 2 shows these errors.

4. Origins of PSN screening errors

Accuracy of the listed methods for PSN linear parameter characterization strongly depends on

- dynamics of patient vertical posture;
- patient breath.

These factors randomly affect the spatial position of the capture device toward the patient skin surface, leading to the random distortions of captured images scale, and, as a consequence, preventing accurate characterization of PSN linear parameters. In photogrammetric theory [34] mathematical apparatus for estimation of image scale dependence on spatial position of capture device toward examined object exists; it is possible to determine the random error of measurements associated with images scale fluctuations utilizing this theoretical approach. Image scale distortions, associated with angular deviation (θ) of camera spatial position toward the object, are described with the following equation:

$$F(k_\theta, c_\theta) = k_\theta^2 \max_\varphi \left[(k_\theta \cdot \cos(\varphi) + c_\theta \cdot \sin(\varphi))^2 + \sin(\varphi)^2 \right]^{-\frac{1}{2}} - 1 , \quad (1)$$

where

$$k_\theta = \cos(\theta) - \frac{x}{f}\sin(\theta) , \quad c_\theta = \frac{y}{f}\sin(\theta) .$$

Table 3 shows the values of maximal spatial deviation of patient's body toward the capture device and corresponding measurement error. Table 4 shows the maximal spatial deviation of the patient's body surface toward the capture device when breathing and corresponding measurement errors. We obtain these errors for the typical experimental conditions: f = 21.6 mm is a focal length the capture device lens, x = y = d/2 = 4.5 mm are the image sizes, d is a length of the image, Z=1800 mm is a distance between capture device and patient. Table 5 shows the total error for linear and area measurements.

*** Formatting Issue - Best Available Paper/Graphic ***

Table 3. Measurement errors caused by vertical posture maintenance.

Factor	Maximal value	Scale changes	Measurement error, %
1. Angle changing by body rotation in the object plane	$\alpha_x = 0{,}9°$	$\delta^L_{\alpha x} = F(k_{\alpha x}, c_{\alpha x})$ $\delta^S_{\alpha x} = \delta^L_{\alpha x} \cdot \sqrt{2}$	0,3 0,4
2. Angle changing by body rotation to the frontal vertical plane	$\alpha_y = 1{,}2°$	$\delta^L_{\alpha y} = F(k_{\alpha y}, c_{\alpha y})$ $\delta^S_{\alpha y} = \delta^L_{\alpha y} \cdot \sqrt{2}$	0,4 0,5
3. Angle changing by body rotation relative the vertical axis	$\beta=2{,}0°$	$\delta^L_{\beta} = F(k_{\beta}, c_{\beta})$ $\delta^S_{\beta} = \delta^L_{\beta} \cdot \sqrt{2}$	0,6 0,9
4. Distance changing when the body is tilted to the object plane	$\Delta Z_{\alpha x}=24$mm[a]	$\delta^L_{\Delta Z\alpha} \frac{\Delta Z_{\alpha x}}{Z + \Delta Z_{\alpha x}}$ $\delta^S_{\Delta Z\alpha} = \delta^L_{\Delta Z\alpha} \cdot \sqrt{2}$	1,3 1,8
5. Distance changing by body rotation relative the vertical axis	$\Delta Z_{\beta}=7$mm[b]	$\delta^L_{\Delta Z\beta} \frac{\Delta Z_{\beta}}{Z + \Delta Z_{\beta}}$ $\delta^S_{\Delta Z\beta} = \delta^L_{\Delta Z\beta} \cdot \sqrt{2}$	0,4 0,6
Result		$\delta^L_1 = \sqrt{(\delta^L_{\alpha x})^2 + (\delta^L_{\alpha y})^2 + (\delta^L_{\beta})^2 + (\delta^L_{\Delta Z\alpha})^2 + (\delta^L_{\Delta Z\beta})^2}$ $\delta^S_1 = \delta^L_1 \cdot \sqrt{2}$	1,6 2,2

[a] $\Delta Z_{\alpha x} = h \cdot \sin(\alpha_x)$, where h = 1500 mm – position chest of athletic men average stature (1700 mm).

[b] $\Delta Z_{\beta} = \frac{b}{2} \cdot \sin(\beta)$, where b = 391 mm – the average width of the shoulders athletic men.

Table 4. Measurement errors caused by breathing.

Factor	Maximal value	Scale changes	Measurement error, %
1. Distance changing	$\Delta Z=25$mm[a]	$\delta^L_{\Delta Z} \frac{\Delta Z}{Z + \Delta Z}$ $\delta^S_{\Delta Z} = \delta^L_{\Delta Z} \cdot \sqrt{2}$	1,4 1,9
2. Angle changing by surface rotation to the frontal vertical plane	$\alpha = 5{,}0°$ [b]	$\delta^L_{\alpha x} = F(k_{\alpha x}, c_{\alpha x})$ $\delta^S_{\alpha x} = \delta^L_{\alpha x} \cdot \sqrt{2}$	1,8 2,6
Result		$\delta^L_2 = \sqrt{(\delta^L_{\Delta Z})^2 + (\delta^L_{\alpha})^2}$ $\delta^S_2 = \delta^L_2 \cdot \sqrt{2}$	2,3 3,2

[a] the amplitude of the chest wall during breathing [35].

[b] $\alpha = arctg(\frac{\Delta Z}{c})$, where c = 300 mm – the maximum distance from most mobile point of chest to stationary

point during quiet breathing for athletic men average stature (1700 mm).

Table 5. Total measurement error for linear dimensions and area.

Name	Equation	Value, %
Measurement error of linear dimensions due to dynamics of patient vertical posture, δ_1^L	$\delta_1^L = \sqrt{(\delta_{\alpha x}^L)^2 + (\delta_{\alpha y}^L)^2 + (\delta_\beta^L)^2 + (\delta_{\Delta Z\alpha}^L)^2 + (\delta_{\Delta Z\beta}^L)^2}$	1.6
Measurement error of area due to dynamics of patient vertical posture, δ_1^S	$\delta_1^S = \delta_1^L \cdot \sqrt{2}$	2.2
Measurement error of linear dimensions due to patient breathing, δ_2^L	$\delta_2^L = \sqrt{(\delta_{\Delta Z}^L)^2 + (\delta_\alpha^L)^2}$	2.3
Measurement error of area due to patient breathing, δ_2^S	$\delta_2^S = \delta_2^L \cdot \sqrt{2}$	3.2
Total measurement error of linear dimensions, δ_Σ^L	$\delta_\Sigma^L = \sqrt{(\delta_1^L)^2 + (\delta_2^L)^2}$	2.8
Total measurement error of area, δ_Σ^S	$\delta_\Sigma^S = \delta_\Sigma^L \cdot \sqrt{2}$	4

5. Capture device position

We utilized the described criteria for formulating the requirements for accuracy of capture device positioning toward the patient body. Allowable deviation of distance between the camera and the patient is calculated

$$\Delta Z = \frac{\delta_{\Delta Z}^L \cdot Z}{1 - \delta_{\Delta Z}^L}, \qquad (2)$$

where

$$\delta_{\Sigma П}^L = \sqrt{(\delta^L)^2 - (\delta_\Sigma^L)^2}, \quad \delta_{\Sigma П}^L = \sqrt{(\delta_{\Delta Я}^L)^2 + (\delta_\alpha^L)^2 + (\delta_\beta^L)^2}, \quad \delta_i^L = \frac{\delta_{\Sigma \Pi}^L}{\sqrt{3}}.$$

For a typical measurement conditions equation (2) gives ΔZ=77 mm.

Required accuracy of angular positioning of the camera relative to the patient body (allowable elevation and azimuthal disorientation of capture device) is calculated:

$$\Delta\alpha = F(k_\alpha, c_\alpha), \qquad (3)$$

and it corresponds to $\Delta\alpha = 9.2^\circ$ for a typical experiment conditions.

6. Illumination of object plane and color coordinates

Illumination of the object plane could impact the accuracy of color parameters characterization due to the number of factors. We have experimentally estimated an impact of changes in object plane illumination on the color coordinates. The illumination was varied within range 1300 … 500 Lux, test

*** Formatting Issue - Best Available Paper/Graphic ***

object images were recorded, and color coordinates corresponding to different illumination conditions were estimated. Thus, we have constructed regression dependencies based on the set of experimental results, and we have estimated an error for color coordinate measurements depending on illumination conditions. It is shown, that color coordinates of test object do not change within the specified illumination range; the measurement error does not exceed 4%.

7. Conclusion

In this paper we have presented a complex of PSN clinical parameters, which could be determined in an automatic mode. We have proposed the PSN measured parameters, and calculated allowable measurement errors; the origin of these errors is discussed. Required accuracy for capture device positioning toward the patient body have been estimated. The results of the present work have been used for creation of an automated device for PSN screening diagnosis based on PSN imaging and subsequent digital image processing.

Acknowledgments

The work was supported by the grant from Russian Scientific Foundation, Project # 14-15-00758.

References

[1] Trapeznikov N N, Aksel' E M and Barmina N M 1998 *The state of the oncological help to the population of the CIS countries in 1997* (Moscow: Russian oncological scientific center named. N.N. Blokhina RAMS) 193
[2] Shaikh W R, Xiong M, Weinstock M A 1978 to 2007 *Archives of Dermatology* **148**(1) 30-6
[3] Helsing P 1997 *Tidsskr Nor Laegeforen* **117**(21) 3107-08
[4] Magennis D P and Orchard G E 1999 *British Journal of Biomedical Science* **56**(2) 134-44
[5] Konopackova O M and Jandarova L F 1990 *Proc. all-Union Symposium Clinic and treatment of malignant melanoma of the skin* 33–34
[6] Ahmedov B P and Hasanova Z M 2002 *Modern Oncology* **4**(4) 199-200
[7] Chissov V I, Starinsky V V and Petrova G V 2011 *The state of the oncological help to the population of Russia in 2010* (Moscow: FSI P.A. Hertzen Moscow Cancer Research Institute the health Ministry of Russia) 188
[8] Patel J K, Konda S, Perez O A, Amini S, Elgart G and Berman B 2008 *Eur. J. Dermatol.* **18**(6) 617-31
[9] Akhmatova A M, Potekaev N N, Reshetov I V, Tkachenko S B, Micenko A V and L'vov A N 2012 *Clinical Dermatology and Venereology* **10**(2) 32-36
[10] Moncrieff M, Cotton S, Claridge E, Hall P 2002 British Journal of Dermatology **146** 448-57
[11] Zonios G and Dimou A 2009 *Opt. Exp.* **17**(3) 1256-67
[12] Claridge E, Cotton S, Hall P and Moncrieff M 2003 *Medical Image Analysis* **7** 489-502
[13] Rubins U, Zaharans J and Spigulis J 2012 *LJPTS* **49**(5) 4-8
[14] Akhmatova A M, Potekaev N N, Reshetov I V, Matorin O V, Frank, G A and Tkachenko S B 2012 *Oncosurgery* **4**(4) 15-19
[15] Akhmatova A M, Potekaev N N, Reshetov I V, Tkachenko S B and Lukasheva N N 2011 *Clinical dermatology and venereology* **9**(5) 38-45
[16] Scalvenzi M, Lembo S, Francia M G and Balato A 2008 *International Journal of Dermatology* **47**(10) 1015-23
[17] Aspres N, Egerton I B, Lim A C and Shumack S P 2003 *Australas J. Dermatol.* **44**(1) 19–27
[18] Dill-Muller D and Maschke J 2007 *J Dtsch. Dermatol. Ges.* **5**(8) 689-707
[19] Mofid Y, Ossant F, Imberdis C, Josse G and Patat F 2009 *IEEE Trans. Ultrason. Ferroelectr. Freq. Control.* **53**(5) 925-35
[20] Branzan A L, Landthaler M and Szeimies R M 2007 *Lasers Med. Sci.* **22**(2) 73–82
[21] Abramovits W and Stevenson L C 2003 *Clin. Dermatol.* **21**(5) 353-58
[22] Izatt J A, Kulkarni M D, Yazdanfar S, Barton J K, Welch A J 1997 *Opt. Lett.* **22**(18) 1439-41

*** Formatting Issue - Best Available Paper/Graphic ***

[23] Mogensen M, Thrane L, Jorgensen T M, Andersen P E and Jemec G B E 2009 *J. Biophotonics* **2**(6-7) 442-51
[24] Reshetov I V, Potekaev N N, Arutyunyat L S, Zaletaev D V, Matorin O V and Kudrin K G 2009 *Russian oncological journal* **5** 54-56
[25] Zaytsev K I et al 2014 *Journal of Physics: Conference Series* **486**(1) 012014
[26] Sokolov D V, Vorojcov G N, Mahson A N et al 2008 *Russian Oncological J.* **4** 6-10
[27] Kudrin K G 2014 *Oncosurgery* **6**(1) 71-76
[28] Klaus W and Richard J 2009 *Fitzpatrick's Color Atlas and Synopsis of Clinical Dermatology: Sixth Edition* (McGraw-Hill Professional)
[29] Spiridonov I N, Kudrin K G, Reshetov I V, Matorin OV 2010 *Modern problems of dermatology, immunology and medical cosmetology* **1**(8) 28-33
[30] Potekaev N N 2010 *Dermatoscopy in clinical practice: a guide for doctors* (Moscow: Publishing Group MDV)
[31] Haratishvili T K , Belysheva T S, Vishnevskaya YaV, Kolobyakov A A and Aliev M D 2010 *Modern problems of dermatology, immunology and medical cosmetology* **2**(2) 5-14
[32] Chissov V I, Starinskii V V, Aleksandrova L M et al 2013 *Early detection and prophylaxis of skin melanoma* (Moscow: FSBI P.A. Hertzen Moscow Cancer Research Institute the Ministry of Health of the Russia)
[33] Potekaev N, Reshetov I, Shuginina E, Arutyunyan L, Matorin O, Kuz'mina T 2009 *Doctor* **1** 2-4
[34] Fairchild M D 2013 *Color Appearance Models, Third Edition* (Chichester: Wiley-IS&T Series in Imaging Science and Technology)
[35] Agapov S V 1996 *Photogrammetry Pushbroom* (Moscow: Kartgeocentr – Geoizdat)
[36] Edited by A.S. Bugaeva, S.I. Ivashova, I.Ya. Immoreeva 2010 *Bioradiolocation* (Moscow: BMSTU)

Structural light focusing phenomenon and enhanced second harmonic generation in $NaNO_2$-infiltrated opal photonic crystal

Kirill I Zaytsev*, Gleb M Katyba, Egor V Yakovlev, Ismail N Aliev, Alexey M Khorokhorov and Stanislav O Yurchenko**

Bauman Moscow State Technical University, 105005, 2nd Baumanskaya str. 5, Moscow, Russia

E-mail: *kirzay@gmail.com, **st.yurchenko@mail.ru

Abstract. We report new experimental results on enhanced second harmonic generation using a structural light focusing phenomenon in photonic crystals (PCs). We use opal-based PC, infiltrated with $NaNO_2$ and pumped with femtosecond laser pulses at various incidence angles, in order to examine the dependence of second harmonic generation efficiency on the pumping wavelength location toward the PC band-gap. We demonstrate one order enhancement of second harmonic generation in case of PC band-gap pumping in comparison to non-band- gap pumping. Second harmonic generation is performed in reflection mode with the maximum of generation in the direction of mirror reflection. We demonstrate that the spectrum of second harmonic does not narrow with the quasi-phase matching condition in case of band-gap generation, and second harmonic spectrum corresponding to non-band-gap generation undergoes 1.5 times narrowing due to the quasi-phase matching.

1. Introduction

The problem of highly efficient high optical harmonic generation are of high importance in photonics, nonlinear optics, and laser physics. Photonic band-gap (BG) structures, photonic crystals (PC) and quasi-crystals, metamaterials have attracted considerable interest as a possible media for this purpose [1–12]. The PC is a structure characterized by a spatially periodic distribution of medium optical properties, and this feature of PCs leads to appearance of numerous physical effects, including nonlinearity of dispersion relation [1–4] and optical field localization [10–12]. PCs are novel media for nonlinear optics and laser physics.

Many papers are dedicated to enhancement of nonlinear light conversion in a PC on the basis of fulfillment of the quasi-phase matching condition utilizing nonlinearity of PC dispersion relation $\mathbf{q}(\omega)$. In paper [6] the condition for enhance sum-frequency generation is examined by means of tuning the spectral position of the two mixing wavelengths near the PC band-gap. In manuscripts [7–9] enhanced nonlinear conversion is produced by compensation of the phase mismatches between the fundamental and the second (or third)-harmonic waves, when the harmonic wavelength was tuned across the band-gap of the colloidal PC, $\lambda_{2\omega/3\omega} \sim \lambda_{BG}$, and the fundamental wavelength was far red-shifted, $\lambda_\omega \gg \lambda_{BG}$. Papers [10-12] introduces another mechanism for nonlinear conversion enhancement in PSs based on structural light focusing effect, which leads to strong optical field localization near the PC surface in case of band-gap pumping. Problem of experimental observation and theoretical description of nonlinear

Content from this work may be used under the terms of the Creative Commons Attribution 3.0 licence. Any further distribution of this work must maintain attribution to the author(s) and the title of the work, journal citation and DOI.
Published under licence by IOP Publishing Ltd

conversion enhancement via the structural light focusing in PCs and quasi-PCs [13] is of high importance for the development of novel nonlinear media.

This paper presents novel experimental results of studying the second harmonic generation enhancement via the structural light focusing phenomenon in PCs. Second harmonic generation is experimentally observed in $NaNO_2$-infiltrated opal PC pumped with the femtosecond optical pulses, and the dependence of second harmonic intensity on incidence and reflection angles, $\theta_I = \theta_R$, and, as a consequence, on fundamental pumping wavelength, λ_ω, position towards the PC band-gap, λ_{BG}, is studied. The highest second harmonic intensity corresponds to $\lambda_\omega = \lambda_{BG}$, and almost one-order enhancement of the second harmonic generation is experimentally achieved in this case in comparison to non-band-gap PC pumping, $\lambda_\omega \gg \lambda_{BG}$. The physical origin of observed phenomenon is associated with structural light focusing effect, leading to strong optical field localization and highly efficient PC pumping.

2. Experimental results

We have used opal globular PC, consisting of fused silica microspheres having $d \approx 400$ nm diameter, and made of colloid solution, as a basis for nonlinear PC manufacturing. Fused silica microspheres were close-packed in a face-centered cubic (FCC) lattice, and the voids between globules were infiltrated with melted $NaNO_2$, having lower melting temperature then the fused silica. Figure 1 shows the schematic representation of the PC and the process of nonlinear light scattering. Poly-domain $NaNO_2$ crystalline structures, possessing non-zero quadratic susceptibility $\widehat{\chi}^{(2)} \neq 0$ [14] were formed in the voids after the PC cooling. Sample manufacturing and PC void full-filling were controlled by means of scanning electron microscopy. Fused silica globules in (1 1 1) PC layer possess regular hexagonal arrangement with 0.9 factor of void full-filling. The (1 1 1) layers of the PC are periodic along the growing direction, thus, the light is assumed propagating along or close to the [1 1 1] direction of the structure. The result of the PC band-gap characterization is presented in the figure 2 and corresponds to the incidence and reflection angles $\theta_I = \theta_R = 10^0$: sharp band-gap centered at $\lambda_{BG} = 980$ nm exists.

We have utilized the femtosecond Yb:KGW laser with a linear polarized output to pump the sample at various angles of incidence θ_I. Fundamental wavelength of femtosecond laser was $\lambda_\omega = 1026$ nm; pulse duration was $\tau \simeq 250$ nm; pulse repetition rate was $f = 200$ kHz; and average power was $3,6$ W. Second harmonic polarization was controlled with a dichroic

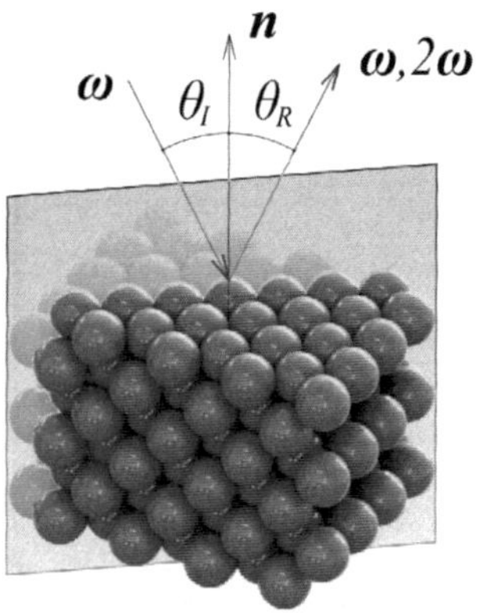

Figure 1. Schematic representation of nonlinear light scattering on a PC. PC filler is not presented in a scheme for simplicity.

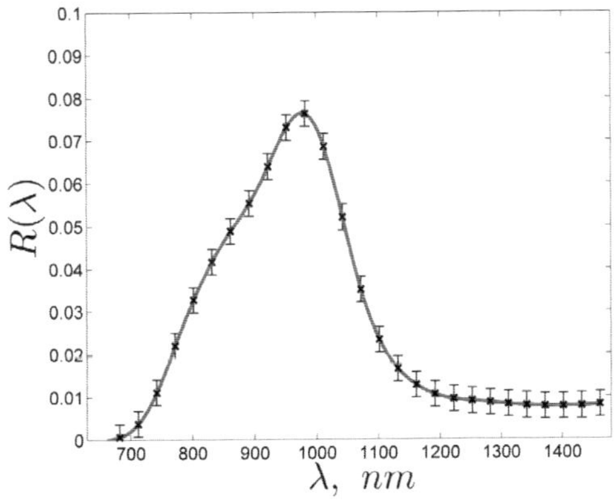

Figure 2. Reflectivity of $NaNO_2$-infiltrated opal PC corresponding to the incidence and reflection angles $\theta_I = \theta_R = 10°$. The peak at this curve corresponds to PC band-gap, $\lambda_{BG} = 980$ nm.

film polarizer to be orthogonal to the plane of incidence. Therefore, s-polarized fundamental wave radiated the sample, and s-polarized second harmonic wave was detected. Laser beam was slightly focused on the PC surface, thus, $D = 2$ mm beam spot diameter was achieved. After nonlinear light scattering the spectrum of the s-polarized optical field intensity was detected at far-field utilizing the diffraction grating spectrometer and the optical fiber probe. Since both the sample and the probe were placed on motorized rotational stages, incidence and reflection angles could be varied, and nonlinear scattering indicatrix could be measured. The fiber probe is used in order to maximize the resolution of the SHG indicatrix characterization.

Figure 3 shows the spectrum of the s-polarized second harmonic corresponding to the incidence and reflection angles $\theta_I = \theta_R = 10°$. In this case the band-gap pumping condition, $\lambda_\omega = \lambda_{BG}$, is satisfied, and highest second harmonic intensity is observed. It is in compleat agreement with the results of numerical simulations performed in [11]. In [11] the enhancement of second harmonic generation was predicted for the case of pumping wavelength matches the PC band-gaps

$$\lambda_\omega = \lambda_{BGi}, \quad i = 1, 2, ..., N. \tag{1}$$

Almost 2% conversion efficiency was reached for the band-gap pumping: ≈ 70 mW power of s-polarized second harmonic, radiated in the direction of mirror reflection and having approximately $\pm 10°$ divergence, was measured utilizing the power meter.

With an increase of $\theta_I = \theta_R$ the PC band-gap, λ_{BG}, undergoes blue–shift [9,12]

$$\lambda_{BG} \cong 2d\sqrt{\tilde{n}^2 - \sin^2(\theta)}, \tag{2}$$

where $\tilde{n}$ is an effective refractive index of PC structure

$$\tilde{n} = \sqrt{(1-f)\, n_{SiO_2}^2 + f n_{NaNO_2}^2}, \tag{3}$$

n_{SiO_2} is a refraction index of fused silica, n_{NaNO_2} is a refraction index of $NaNO_2$ voids, and f is $NaNO_2$ volume fraction. Therefore, the dependence of second harmonic generation efficiency on the fundamental wavelength location, λ_ω, toward the band-gap, λ_{BG}, could be examined by rotating the PC. Figure 4 shows the decreasing of the second harmonic generation intensity with growth of equal pair of incidence and reflection angles, $\theta_I = \theta_R$. One order enhancement of the

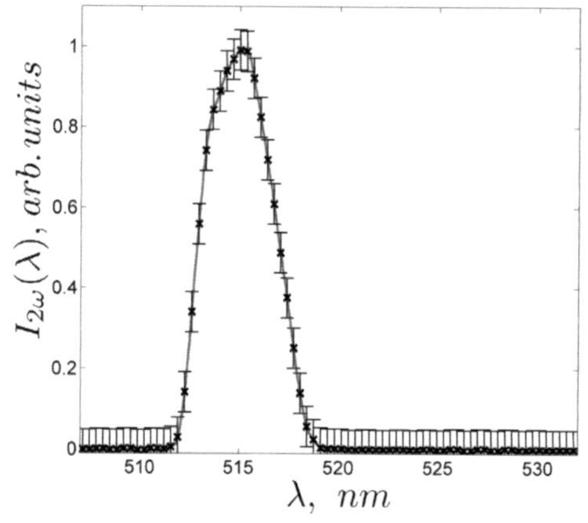

Figure 3. Spectrum of second harmonic generated in $NaNO_2$-infiltrated opal PC in case of optimal incidence and reflection angles, $\theta_I = \theta_R = 10°$.

second harmonic generation was experimentally observed for the band-gap pumping, $\lambda_\omega = \lambda_{BG}$ ($\theta_I = \theta_R = 10°$), in comparison to non-band-gap pumping, $\lambda_\omega > \lambda_{BG}$, ($\theta_I = \theta_R > 10°$).

Figure 5 shows the dependence of second harmonic spectrum width at full-width at half maximum, $FWHM_{2\omega}$, towards the increasing angles, $\theta_I = \theta_R$. Strong decrease of $FWHM_{2\omega}$ is observed with $\theta_I = \theta_R$ increasing, and, as a consequence, with λ_{BG} blueshifting. This curve confirms, that nonlinear generation occurs in the thin near-surface area of the PC in case of the band-gap pumping, $\lambda_\omega = \lambda_{BG}$ [11,12], and it performs in the PC volume in opposite case, $\lambda_\omega \neq \lambda_{BG}$ [7–9]. This narrowing of second harmonic spectrum is associated with the quasi-phase

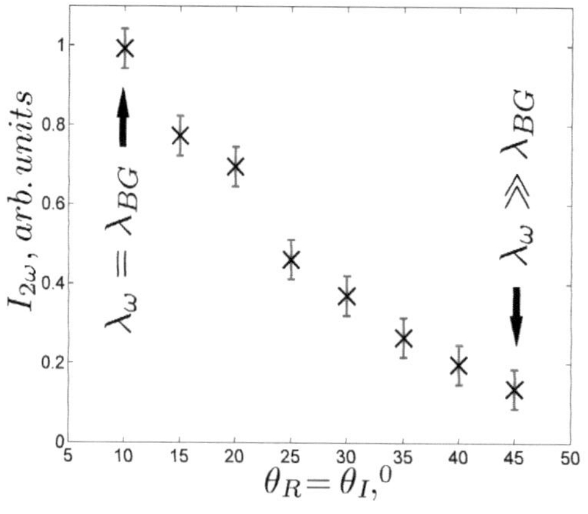

Figure 4. Dependence of second harmonic intensity, $I_{2\omega}$, on $\theta_I = \theta_R$. Varying $\theta_I = \theta_R$ we are examining the dependance of second harmonic generation efficiency on the ration $\lambda_\omega/\lambda_{BG}$, and the most efficient second harmonic generation corresponds to PC band-gap pumping, $\theta_I = \theta_R = 10°$ and $\lambda_\omega = \lambda_{BG}$.

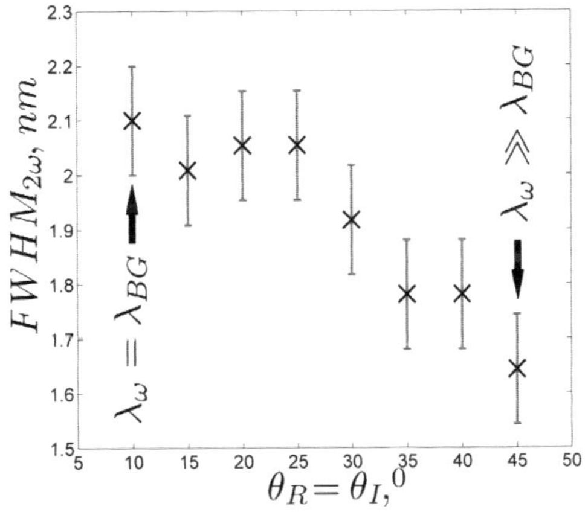

Figure 5. Dependence of second harmonic spectrum width, $FWHM_{2\omega}$, on $\theta_I = \theta_R$. The most wideband second harmonic generation corresponds to PC band-gap pumping, $\theta_I = \theta_R = 10°$ and $\lambda_\omega = \lambda_{BG}$, since no quasi-phase matching condition [7–9] should be satisfied for the band-gap generation case [12].

matching condition, which should be satisfied in case of generation occurs in PC volume [7–9].

One could perform efficient second harmonic generation in PC volume if the condition of effective nonlinear diffraction [7-9,12] is satisfied

$$\mathbf{k}_{2\omega} = \mathbf{k}_\omega + n\mathbf{G}_{111} \tag{4}$$

where $\mathbf{k}_\omega$ is a fundamental wave vector, $\mathbf{k}_{2\omega}$ is a second harmonic wave vector, $\mathbf{G}_{111}$ is a reciprocal vector of the FCC lattice in [1 1 1] direction, and n is an order of the PC band-gap. This condition takes into account interference between the SH fields from all $NaNO_2$ infiltrated voids, and the interference is constructive only for thin bandwidth of the second harmonic spectrum. In case of PC band-gap pumping, $\lambda_\omega = \lambda_{BG}$, second harmonic generation is performed in several near-surface layers of PC [10–12], and condition (4) becomes insignificant, since second harmonic field at far-field region is a superposition of the second harmonic fields generated in the $NaNO_2$-infiltrated voids of several near-surface PC layers. Thus, no narrowing of second harmonic spectrum is observed for the band-gap generation [11,12] in contrast to the previous quasi-phase matching approaches [7–9]. Strong optical field localization due to the structural light focusing and no spectrum narrowing are main advantages of considered technique of second harmonic generation in a PC.

3. Discussions

Observed second harmonic generation demonstrates both an existence of second-order nonlinearity ($\hat{\chi}^{(2)} \neq 0$) in particular PC and the ability for enhanced second harmonic generation due to the structural light focusing phenomenon. It is possible to realize other nonlinear conversions related to second order nonlinearity, such as, parametric conversion, low-frequency wave generation via the optical rectification and parametric decay. The latter nonlinear processes could become a basis for development of novel terahertz devices (sources, detectors, and modulators) [15–17], and these devices could be used for the the needs of terahertz spectroscopy and terahertz imaging [18-26].

Structural light focusing should be inherent to the media possessing third order nonlinearity ($\widehat{\chi}^{(3)} \neq 0$), and, thus, demonstrating the enhancement of third-order nonlinear optical processes, in particular, enhancement of third harmonic generation, is an important problem for future research. Empty 3D opal PCs made of fused silica or polymer globules [27,28] or 2D PCs based on cotton-growth ZnO nanorods [29,30] could become appropriate media for demonstrating third harmonic generation enhancement. Studying the nonlinear effect in quasi-PC structures and partially-ordered media [31-33,13] is another perspective topic for fundamental and applied research, since the structural light focusing could also appear in this structures leading to nonlinear effect enhancement, and to highly efficient interactions of light with matter.

4. Conclusion

In this paper we have reported new experimental results on enhanced second harmonic generation in PC based on the structural light focusing phenomenon. We use opal-based PC, infiltrated with $NaNO_2$ and pumped with femtosecond laser pulses at various angles of incidence, in order to examine the dependance of second harmonic generation efficiency on the spectral position of pumping wavelength toward the PC band-gap. One order enhancement of second harmonic generation was observed in case of band-gap pumping comparing to non-band-gap pumping of the PC. In case of band-gap pumping the maximum of second harmonic generation corresponds to the direction of mirror reflection and has $\pm 10^\circ$ divergence.

5. Acknowledgments

This work was supported by the grant of Russian Scientific Foundation, Project # 14-29-00277.

References

[1] Bykov V P 1972 *J. Experimental and Theoretical Physics* **35** 269–73
[2] Yablonovich E 1987 *Phys. Rev. Lett.* **58** 2059–62
[3] John S 1987 *Phys. Rev. Lett.* **58** 2486–89
[4] Meade R D, Rappe A M, Brommer K D, Joannopoulos J D, Alerhand O L 1993 *Phys. Rev. B* **48** 8434–37
[5] Soljacic M, Joannopoulos J D 2004 *Nature Materials* **3** 211-19
[6] Balakin A V, Bushuev V A, Mantsyzov B I, Ozheredov I A, Petrov E V, Shkurinov A P, Masselin P and Mouret G 2001 *Phys. Rev. E* **63** 046609
[7] Martorell J, Vilaseca R, Corbalan R 1997 *Phys. Rev. A* **55** 4520–25
[8] Markowicz P P, Tiryaki H, Pudavar H, Prasad P N, Lepeshkin N N, Boyd R W 2004 *Phys. Rev. Lett.* **92** 083903
[9] Fedyanin A A, Aktsipetrov O A, Kurdyukov D A, Golubev V G, Inoue M 2005 *Appl. Phys. Lett.* **87** 151111
[10] Zaytsev K I, Gorelik V S, Khorokhorov A M and Yurchenko S O 2014 *J. Phys. Conf. Ser.* **486** 012003
[11] Zaytsev K I, Katyba G M, Yakovlev E V, Gorelik V S and Yurchenko S O 2014 *J. Appl. Phys.* **115** 213505
[12] Zaytsev K I, Yurchenko S O 2014 *Appl. Phys. Lett.* **105** 051902
[13] Yurchenko S O 2014 *J. Chem. Phys.* **140** 134502
[14] Martienssen W, Warlimont H 2005 *Springer Handbook of Condenced Matter and Materials Data* (Berlin: Springer Berlin Heidelberg)
[15] Svintsov D, Vyurkov V, Yurchenko S, Otsuji T, Ryzhii V 2012 *J. Appl. Phys.* **111** 083715
[16] Ryzhii1 V, Otsuji T, Ryzhii M, Leiman V G, Yurchenko S O, Mitin V, Shur M S 2012 *J. Appl. Phys.* **112** 104507
[17] Ryzhii1 V, Otsuji T, Ryzhii M, Ryabova N, Yurchenko S O, Mitin V, Shur M S 2013 *J. Phys. D: Appl. Phys.* **46** 065102
[18] Borodin A V, Panov N A, Kosareva O G, Andreeva V A, Esaulkov M N, Makarov V A, Shkurinov A P, Chin S L, Zhang X-C 2013 *Opt. Lett.* **38** 1906–08
[19] Borodin A V, Esaulkov M N, Frolov A A, Shkurinov A P, Panchenko V Y 2014 *Opt. Lett.* **39** 4092–95
[20] Zaytsev K I, Karasik V E, Fokina I N, Alekhnovich V I 2013 *Opt. Eng.* **52** 068203
[21] Zaytsev K I, Gavdush A A, Karasik V E, Alekhnovich V I, Nosov P A, Lazarev V A, Reshetov I V, Yurchenko S O 2014 *J. Appl. Phys.* **115** 193105
[22] Zaytsev K I, Chernomyrdin N V, Alekhnovich V I 2014 *J. Phys. Conf. Ser.* **486** 012010
[23] Zaytsev K I, Gavdush A A, Lebedev S P, Yurchenko S O 2014 *J. Phys. Conf. Ser.* **486** 012018

[24] Zaytsev K I, Kudrin K G, Koroleva S A, Fokina I N, Volodarskaya S I, Novitskaya E V, Perov A N, Karasik V E, Yurchenko S O 2014 *J. Phys. Conf. Ser.* **486** 012014
[25] Zaytsev K I, Chernomyrdin N V, Gorevoy A V, Trofimov N E, Fokina I N, Alekhnovich V I, Karasik V E, Yurchenko S O 2014 *J. Phys. Conf. Ser.* **486** 012034
[26] Perov A N, Zaytsev K I, Fokina I N, Karasik V E, Yakovlev E V, Yurchenko S O 2014 *J. Phys. Conf. Ser.* **486** 012027
[27] Blanco A, Chomski E, Grabtchak S, Ibisate M, John S, Leonard S W, Lopez C, Meseguer F, Miguez H, Mondia J P, Ozin G A, Toader O, van Driel H M 2000 *Nature* **405** 437–40
[28] Pursiainen O L J, Baumberg J J, Winkler H, Viel B, Spahn P, Ruhl T 2007 *Opt. Exp.* **15** 9553–61
[29] Al-Hilli S M, Al-Mofarji R T, Klason P, Willander M, Gutman N, Saar A 2008 *J. Appl. Phys.* **103** 014302
[30] Israr M Q, Sadaf J R, Yang L L, Nur O, Willander M, Palisaitis J, Persson P O A 2009 *Appl. Phys. Lett.* **95** 073114
[31] Vardeny Z V, Nahata A, Agrawal A 2013 *Nature Photonics* **7** 177-87
[32] Bunkin N F, Yurchenko S O, Suyazov N V, Shkirin A V 2012 *J. Biol. Phys.* **38** 121-52
[33] Bunkin N F, Yurchenko S O, Suyazov N V, Starosvetskiy A V, Shkirin A V, Kozlov V A 2011 *Modeling the cluster structure of dissolved air nanobubbles in liquid media (Book Chapter)* (New-York: Nova Science Publishers Inc.)

Observation of cladding modes spatio-spectral distribution in large mode area photonic crystal fiber

Stanislav O Leonov*[1], Vladimir A Lazarev[1], Alina V Borisova[1], Vladimir V Demidov[2], Andrey D Pryamikov[3]
[1] Bauman Moscow State Technical University, 5/1 2nd Baumanskaya str., Moscow, 105005, Russia
[2] S.I. Vavilov State Optical Institute, Babushkin str., 36/1, St. Petersburg, 192171 Russia
[3] Fiber Optics Research Center, Russian Academy of Sciences, 38, Vavilov str., Moscow, Russia

E-mail: *leonov-st@ya.ru

Abstract. We report the observation of spatio-spectral distribution in cladding modes of a single-mode large mode area photonic crystal fiber. The cladding modes excitation was achieved without any external fiber exposure. The optical field patterns of the cladding modes within different pump wavelength are investigated. To the best of knowledge the spatio-spectral distribution in cladding modes of large mode photonic crystal fiber is demonstrated for the first time. The results are of immediate interest in applications demanding devices based on core and cladding mode coupling in photonic crystal fibers.

1. Introduction

Photonic crystal fibers (PCF) [1, 2] are of particular interest for a number of applications like gas analysis [3], optical coherence tomography [4], supercontinuum sources [5-8], optical frequency metrology [9-11] etc. To date, much effort has been done on understanding and research of PCFs core modes [12-16] and very little activity is observed in the study of cladding modes. Since working principle of many devices based on standard fibers and PCFs involves cladding modes, it is expected that these modes can also play an important role in the development of novel fiber devices based on PCF. Thus, cladding modes play a crucial role in fiber-optic sensing [17-19], in devices based on core and cladding mode coupling [20]. The number of methods was applied to research cladding modes in PCFs. In most cases a cladding mode excitation was made by external fiber exposure. In this respect, in [21] a coupling by a Bragg grating was used to investigate higher-order leaky modes. In [22] excitation of cladding modes by flexural acoustic waves was reported. Coupling between fundamental mode and cladding modes was realized by long period gratings written in pure silica PCFs [23].

In this paper, we observe excitation of cladding modes in a single-mode large mode area (LMA) PCF without any external fiber exposure (acoustic wave or long-period grating) and demonstrate spatial spectral shaping of cladding modes. These results could play an important part in theoretical problems of light guidance and nonlinear effects in photonic crystal fibers.

Content from this work may be used under the terms of the Creative Commons Attribution 3.0 licence. Any further distribution of this work must maintain attribution to the author(s) and the title of the work, journal citation and DOI.
Published under licence by IOP Publishing Ltd

2. Experiments

To understand the cladding modes excitation phenomena in a large mode area PCF we used the fiber with the core diameter 20 μm manufactured by a stack-and-draw method [24]. A microscope image of the end face of the fiber used in experiment is shown in figure 1. The fiber supports a single transverse mode for any wavelength, with an 13.2 μm pitch (*Λ*- center-to-center distance between holes) and an 6 μm hole diameter (*d*). The outer diameter of the fiber (*R*) was 160 μm. The structure of the fiber has four layers of air holes arranged in a hexagonal lattice surrounding the silica core.

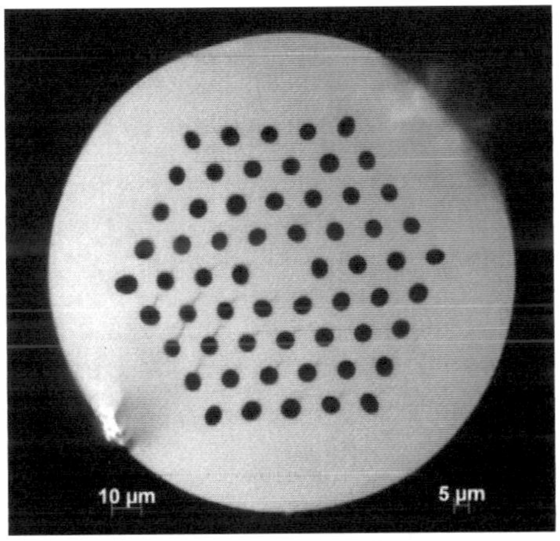

Figure 1. Optical microscope image of the cleaved endface of the PCF. The hole diameter is 6 μm, the pitch is 13.2 μm, the outer diameter is 160 μm.

The refractive indices used for silica and air are 1.45 and 1, respectively. The normalized air hole diameter (d/Λ), and the normalized outer diameters ($2R/\Lambda$) are d/Λ = 0.45, and $2R/\Lambda$ = 12.1, respectively. The PCF with these values was calculated to be a single-mode. The characteristics of the LMA PCF used in the experiments are summarized in Table 1.

Table 1. Characteristics of the LMA PCF.

Core Diameter (μm)	Λ (μm)	d/Λ	Fiber Length (m)	ZDW (nm)
20	13.2	0.45	1	1250

An experimental setup is illustrated in figure 2. A laser source is employed by supercontinuum source (SC-400, Fianium Ltd.) offers a wavelength range of 390 nm extending to 2600 nm with 250 nJ pulse energy and total power of 4 W. The laser operates at repletion rate of 40 kHz and produces pulses with temporal width of ~6 ps. Laser radiation of SC-400 passed through acousto-optic filter (AOTF-FS, Fianium Ltd.).

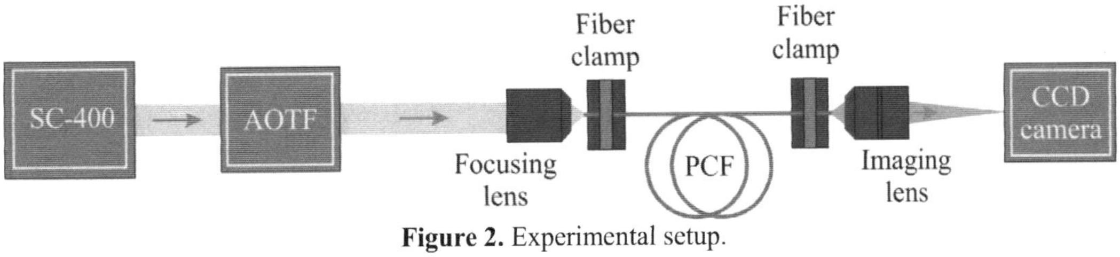

Figure 2. Experimental setup.

AOTF-FS is fitted with a visible AOTF crystal (400-650 nm) and enables up to 8 simultaneous tunable spectral channels with bandwidth of 2-4 nm to be selected from any SC-400 supercontinuum spectrum. The output radiation from AOTF was focused by a 40X microscope focusing lens into the PCF under test, which was mounted in a special three-dimensional positioning table (MAX361D/M, Thorlabs). To observe the near-field pattern distribution of the fiber output radiation, the transmitted light from the output edge of PCF was detected by a CCD camera (DCU224C, Thorlabs) through 40X microscope objective which acts as an imaging lens.

In our experiment cladding modes excitation was observed by passing radiation from two different AOTF spectral channels through the PCF simultaneously. Obtained near-field images of light transmitted through the fiber is shown in figure 3. Near-field image in figure 3 (a) was obtained when AOTF-FS was tuned to two spectral channels at 532 nm and 600 nm simultaneously, in figure 3 (b) at 532 nm and 650 nm and in figure 3 (c) at 485 nm and 690 nm respectively.

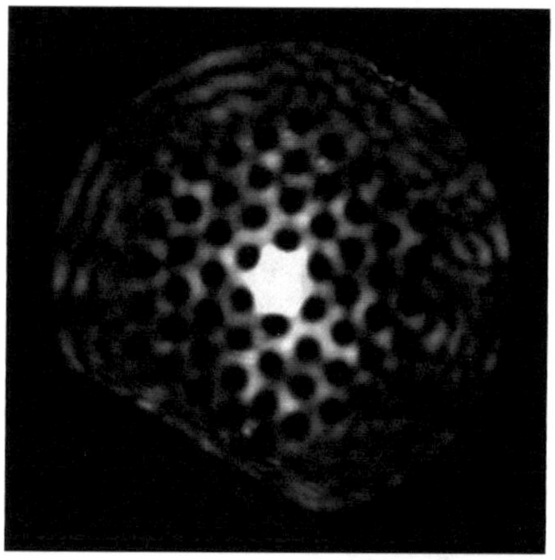

a

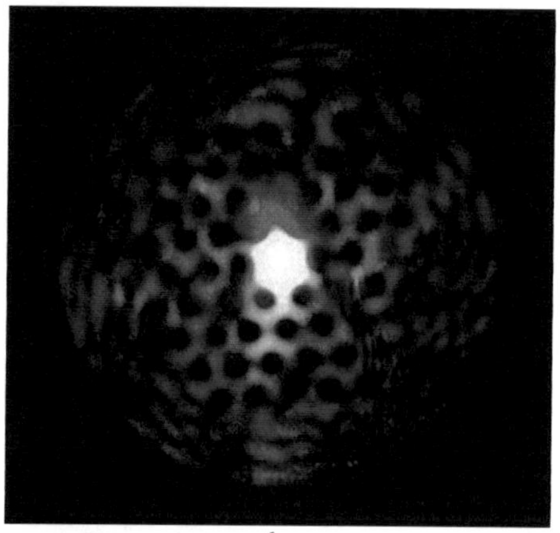

b

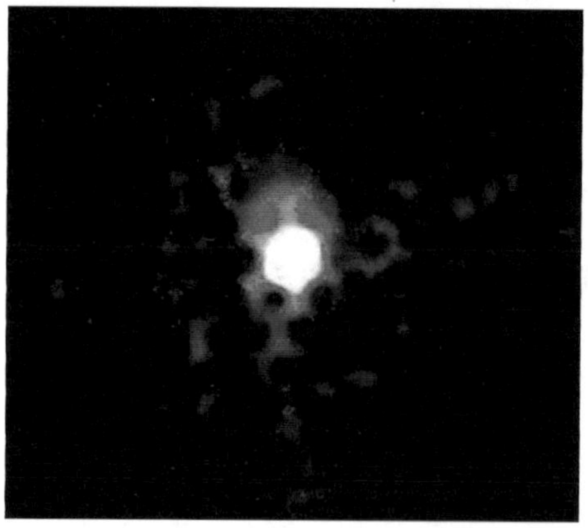

c

Figure 3. The near-field images of light transmitted through the PCF when acousto-optic filter (AOTF-FS) was tuned to 532 nm and 600 nm simultaneously (a), at 532 nm and 650 nm (b) and at 485 nm and 690 nm (c).

In figure 3 (a) and (b) the light has spread out into the cladding area, confirming the excitation of the cladding modes. In figure 3 (c) light at wavelengths of 485 nm and 690 nm has also spread out into the cladding area but didn't confirm the excitation of the cladding modes. As opposed to excitation mechanisms based on long period gratings [23] or acoustic waves [22] in our case we obtain cladding modes excitation without any external fiber exposure. We assume the excitation of cladding modes is caused by macrobend and confinement losses of PCF that result in effective coupling between core and cladding modes. During the experiment the dependence of coupling efficiency on wavelength range was found. Thus, coupling is effective at wavelengths of 532 nm, 600 nm and 650 nm in comparison with wavelengths of 485 nm and 690 nm. In Fig. 3 cladding mode spatial spectral shaping is also observed. We assume this spatial spectral shaping can be a result of cladding modes interactions and is of interest to further investigations.

3. Conclusions

In summary, we observe, for what is believed to be the first time, the excitation of cladding modes in a single-mode large mode area photonic crystal fiber by supercontinuum source equipped with tunable acousto-optic filter without any external fiber exposure. The optical field patterns of the cladding modes are presented. This experimental result provides essential information for design and development cladding-mode-based application of PCFs, also the results are of interest in devices based on core and cladding mode coupling and spatial spectral distribution.

References

[1] Knight J C, Birks T A, Russell P St J and Atkin D M 1996 *Opt. Lett.* **21** 1547-49
[2] Russell P St J 2003 *Science* **299** 358-62
[3] Buric M P, Chen K P, Falk J, Woodruff S D 2008 *Appl. Opt.* **47** 4255-61
[4] Humbert G, Wadsworth W, Leon-Saval S, Knight J, Birks T, Russell P St J, Lederer M, Kopf D, Wiesauer K, Breuer E, Stifter D 2006 *Opt. Exp.* **14** 1596-1603
[5] Stark S P, Travers J C, Russell P St J 2012 *Opt. Let.* **37** 770-72
[6] Pureur V, Dudley J M 2010 *Opt. Let.* **35** 2813-15
[7] Savitski V G, Yumashev K V, Kalashnikov V L, Shevandin V S, Dukel'skii K V 2007 *Opt. Quant. Electron.* **39** 1297-1309
[8] Pasishnik A S, Leonov S O 2014 *Proc. of SPIE* **9136** 91361F
[9] Zaugg C A, Klenner A, Mangold M, Mayer A S, Link S M, Emaury F, Golling M, Gini E, Saraceno C J, Tilma B W and Keller U 2014 *Opt. Exp.* **22** 16445-55
[10] Jones D J, Diddams S A, Ranka J K, Stentz A, Windeler R S, Hall J L and Cundiff S T 2000 *Science* **288** 635–39
[11] Diddams S A, Jones D J, Ye J, Cundiff S T, Hall J L, Ranka J K, Windeler R S, Holzwarth R, Udem T and Hänsch T W 2000 *Phys. Rev. Lett.* **84** 5102–05
[12] Trabold B M, Novoa D, Abdolvand A, Russell P St J 2000 *Opt. Let.* **39** 3736-39
[13] Trabold B M, Abdolvand A, Euser T G, Walser A M, Russell P St J 2013 *Opt. Lett.* **38** 600-2
[14] Trabold B M, Novoa D, Abdolvand A and Russell P St J 2014 *Opt. Lett.* **39** 3736-39
[15] Trabold B M, Abdolvand A, Euser T G and Russell P St J 2013 *Opt. Exp.* **21** 29711-18
[16] Messerly M J, Pax P H, Dawson J W, Beach R J and Heebner J E 2013 *Opt. Exp.* **21** 12683-90
[17] Wu D K C, Kuhlmey B T and Eggleton B J 2009 *Opt. Lett.* **34** 322-24
[18] Ritari T, Tuominen J and Ludvigsen H 2014 *Appl. Opt.* **53** 3668-72
[19] Ritari T, Tuominen J and Ludvigsen H 2004 *Opt. Exp.* **12** 4080-87
[20] Qu H, Brastaviceanu T, Bergeron F, Olesik J, Pavlov I, Ishigure T and Skorobogatiy M 2013 *Appl. Opt.* **52** 6344-49
[21] Eggleton B J, Westbrook P S, Windeler R S, Spaelter S and Strasser T A 1999 *Opt. Lett.* **24** 1460-62
[22] Diez A, Birks T A, Reeves W H, Mangan B J and Russell P St J 2000 *Opt. Lett.* **25** 1499-501
[23] Kakarantzas G, Birks T A and Russell P St J 2002 *Opt. Lett.* **27** 1013-15
[24] Dukel'skii K V, Kondrat'ev Yu N, Khokhlov A V, Shevandin V S, Zheltikov A M, Konorov S O, Serebryannikov E E, Sidorov-Biryukov D A, Fedotov A B, Semenov S L 2005 *J. Opt. Technol.* **72** 548-50

Method to improve the accuracy of the geometrical parameters measurement in stereoscopic AOTF-based spectral imagers

A V Gorevoy[1], **A S Machikhin**[2]
[1]Bauman Moscow State Technical University, 5/1, 2nd Baumanskaya street, Moscow, Russia 105005
[2]Scientific and Technological Center of Unique Instrumentation of Russian Academy of Sciences, 15, Butlerova street, Moscow, Russia 117342

Email: [1]strannik-algor@mail.ru, [2]aalexanderr@mail.ru

Abstract. The problem of non-contact geometrical measurements by the stereoscopic spectral AOTF-based imagers is considered. Major factors contributing to the measurement error are analyzed for the first time. Expressions are derived for area measurement inaccuracy on arbitrary shape surfaces, caused by measurement errors of individual points' 3D coordinates with and without a priori information about surface shape. Verification of the obtained expressions with the real experimental data showed that area measurement error for a complex figure, given by the set of points, is caused mainly by ignoring the fact that these points belong to the surface. It is shown that the use of a priori information about investigated surface shape, which is often available from the design documentation, in many cases would radically improve the accuracy of surface defects area measurement. The presented results can be effectively used for the development of new 3D spectral imagers and modernization of existing ones.

1. Introduction

Currently, one of the major problems in the analysis of materials is the automatic detection, segmentation and classification of the elements with certain physical, chemical, biological or other properties, i.e. to construct a "map of the object's properties". The most of approaches are based either on the image processing using a priori information about the required elements or on the selection of objects with respect to certain physical properties during a particular impact on them. The first approach is applicable only when the properties of the objects are known or already obtained by another instrument. The second approach is ineffective when it is necessary to identify and classify a large number of elements with different properties. That is why the most promising solution to the problem is spectral imaging [[1]]. Allowing the simultaneous registration of the spatial and spectral information, these devices combine the advantages of both approaches. Therefore, spectral imagers are widely used in remote sensing, biomedicine, machine vision and other applications [[2]], [[3]].

It is often necessary (e.g., in a study of moving objects) and sufficient (e.g., if a priori information about the spectrum of the inspected object is available) to register spectral images not in the entire tuning range but only at the particular wavelengths. Such arbitrary spectral tuning significantly reduces the time of registration and data processing. Among the tunable optical spectral filters the most promising are acousto-optic (AOTFs). They provide unique collection of features such as

Content from this work may be used under the terms of the Creative Commons Attribution 3.0 licence. Any further distribution of this work must maintain attribution to the author(s) and the title of the work, journal citation and DOI.
Published under licence by IOP Publishing Ltd

programmability, image transmission with high image quality - insignificant (less than 1%) distortion in the visible spectral range [11], [[5]]. Rather high spectral (up to 0,1 nm) and spatial (up to 1000×1000 dots) resolution, fast spectral tuning (~10 μs), compactness and absence of any moving elements make them a precise and ergonomic analytical tool capable of working even in out-of-lab environment [[6]]-[[8]]. Therefore AOTF-based spectral imagers are widely used in various fields of science and technology, in particular, to determine the three-dimensional (3D) structure of objects [[9]], [[10]], [[11]]. In this case, the use of spectral contrast allows getting more information about the position, shape and size of the volume elements of the object.

One of the methods to restore the spectral 3D structure of objects is stereoscopy [[12]]. Stereoscopic imaging at arbitrary wavelengths may be implemented by the optical scheme containing one AOTF in each channel (see Figure 1a) [[9]]. This device forms two light beams carrying the image of the object at different angles. Then stereoscopic images are registered by digital cameras. There are also other schemes of 3D spectrometers, using only one specialized AO cell (see Figure 1b) [[13]].

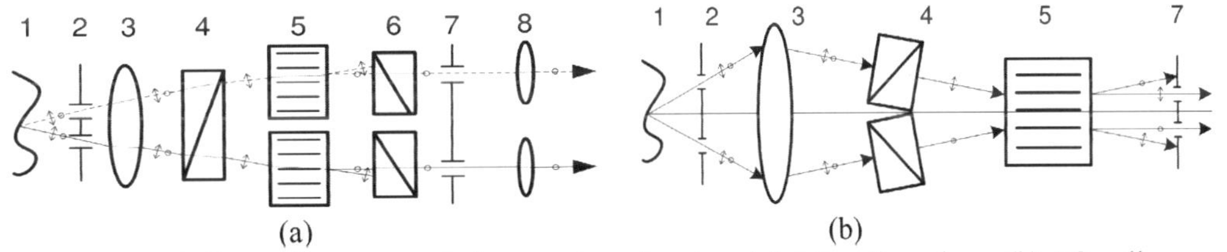

Figure 1. Spectral stereoscopic imagers based on two (a) AO cells and one (b) AO cell
1 – inspected object, 2, 7 – diaphragms; 3, 8 – lenses, 4, 6 - polarizers, 5 – AO cells.
Arrows between optical components show the light polarization direction.

These devices are very effective, for instance, for contrast visualization of the 3D distribution of foreign inclusions or defects in non-destructive testing of various objects (see Figure 2). Usually it is not enough to find and visualize the element, but usually it is also necessary to measure its geometric parameters: length, depth, perimeter, square, etc. Measurement accuracy is highly important as it defines the quality of the whole 3D spectral imaging system.

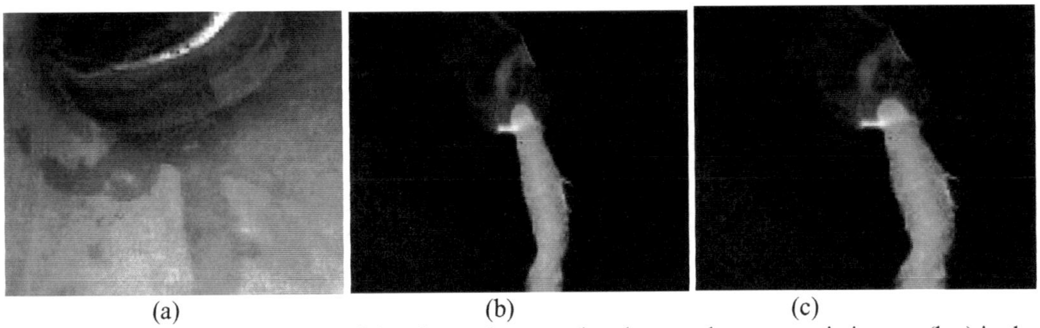

Figure 2. Wide-band image (a) of the pipe and narrow-band spectral stereoscopic images (b,c) in the range $\lambda = 500 \div 503$ nm.

The basic problem of such measurements is that in narrow spectral bands only the boundaries of the detected elements are displayed vividly and not their content (see Figure 2). In this case only 3D coordinates of points on this boundary can be determined accurately. Inside the homogeneous object correlation algorithms for determination of conjugate points lead to large errors. Even when the defect is superficial, i.e. follows the a priori known shape of the surface, the problem has no unique solution due to uncertainty of the imager's orientation relative to the object. Therefore, in such cases, as a rule,

Journal of Physics: Conference Series **584** (2015) 012004 doi:10.1088/1742-6596/584/1/012004

the shape of the object is interpolated by the set of planar triangles with vertices belonging to the boundary [[14]]. If the object has a complicated 3D shape, the error of its area measurement in this case can easily reach 30-50%.

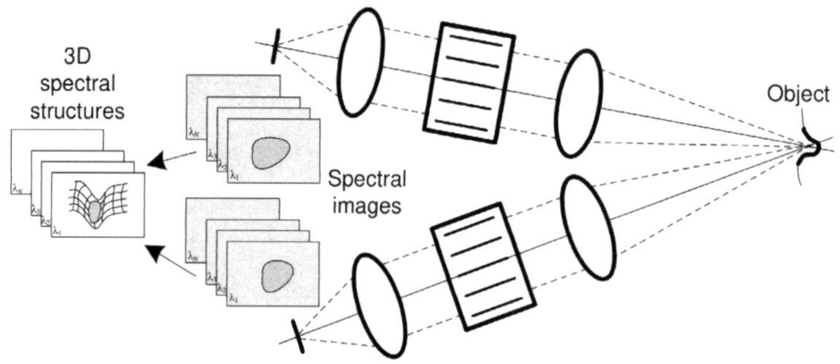

Figure 3. 3D spectral imaging system based on two AOTF-based imagers.

As in the traditional stereoscopic systems, owing to many uncertainties, reliable methods to assess the accuracy of the measurement have not yet been developed. Since the 3D spectral imagers are not only observant but also the measuring devices, the estimation of the measurement error is the key issue in determining the possibility of their application. Therefore, there is a need to develop methods to estimate the accuracy of such stereoscopic measurements.

In this paper, we demonstrate an approach to solving this problem based on the model of the classical two-channel stereoscopic system (see figure 3 and figure 4) and modified algorithms for determining the three-dimensional coordinates of the object points.

2. Camera model and 3D reconstruction algorithm

The registration channels of the stereoscopic system may be represented by the simplified geometric model of the image formation $\mathbf{p}_i = P_i \circ E_i(\mathbf{x}_\mathbf{g})$, where $\mathbf{x}_\mathbf{g}$ and $\mathbf{p}_i$ are 3D coordinates of object points measured in the global coordinate system (GCS) and 2D coordinates of corresponding image points on the image plane of i-th camera; $i = 1..N$, N – number of cameras; E_i is the Euclidean mapping from the GCS to the coordinate system (CS) of i-th camera; P_i is the transformation translating points from the target camera object space onto the image plane (from $A(x_i, y_i, z_i)$ to $a_i(u_i, v_i)$ in the figure 4). We use «$\circ$» notation here to define a composition of transformations, i.e. $P_i \circ E_i(\mathbf{x}_\mathbf{g}) \equiv P_i\left(E_i(\mathbf{x}_\mathbf{g})\right)$.

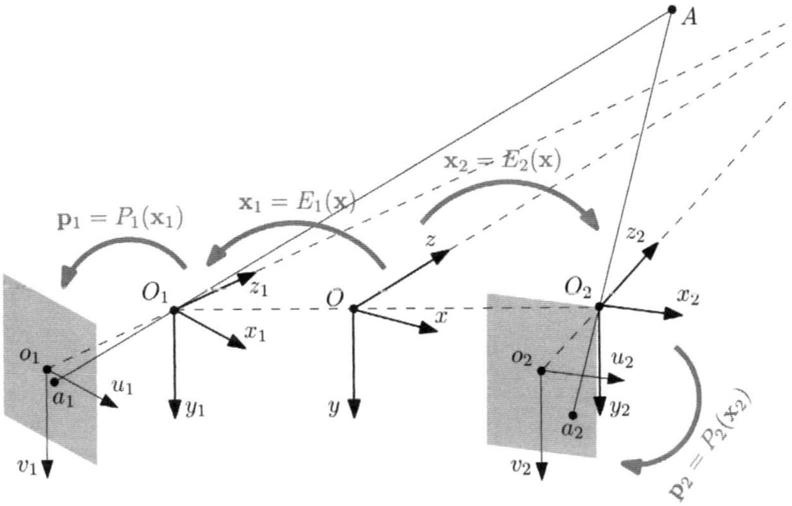

Figure 4. Simplified camera model and two-view geometry.

The mapping P_i determines the unique correspondence $P: L(O,A) \to a(u,v)$ of rays $L(O,A)$ in the object space and points on the image plane. In order to allow 3D reconstruction by means of triangulation, camera model should be invertible and should provide the back-projection transformation $P^{-1}: a(u,v) \to L(O,A)$. The pinhole camera model, the pinhole model, followed by distortion, equidistant and generic polynomial camera model are commonly used [[15]]-[[17]].

The camera placement in 3D space is described by the set of transformations from one of camera-centered CS to each other. Let one of these CS be the GCS and define the set $(\mathbf{R}_i, \mathbf{t}_i)$ converting point coordinates $\mathbf{x}_i = (x_i, y_i, z_i)^{\mathrm{T}}$ in i-th CS to the global coordinates $\mathbf{x}_{\mathrm{g}}$ via following relation: $\mathbf{x}_i = E_i(\mathbf{x}_{\mathrm{g}}) = \mathbf{R}_i\mathbf{x}_{\mathrm{g}} + \mathbf{t}_i$. Any transformation from i-th to j-th CS may be written as [[18]], [[19]]

$$\mathbf{R}_{i\to j} = \mathbf{R}_j\mathbf{R}_i^{-1}, \ \ \mathbf{t}_{i\to j} = \mathbf{t}_j - \mathbf{R}_j\mathbf{R}_i^{-1}\mathbf{t}_i. \tag{1}$$

The set of transformations P_i and E_i is parametrized by the vector $\mathbf{v} = (\mathbf{v}_1^{\mathrm{T}}, \ldots, \mathbf{v}_N^{\mathrm{T}})^{\mathrm{T}}$ of intrinsic and extrinsic camera parameters, which components are to be determined during system calibration [[15]–[20]].

The reconstruction of 3D coordinates $\mathbf{x}_{\mathrm{g}}$ from N projections $\mathbf{p}_i, i = \overline{1,N}$ may be considered as the ray intersection problem $L_i = E_i^{-1} \circ P_i^{-1}(\mathbf{p}_i)$. Because of the uncertainties in corresponding points coordinates $\mathbf{p}_i$, these rays are skew, and the triangulation algorithm T is used for estimation of $\hat{\mathbf{x}}_{\mathrm{g}}$ by minimizing a functional C, i.e.

$$\hat{\mathbf{x}}_{\mathrm{g}} = T(\mathbf{p}, \mathbf{v}) = \underset{\hat{\mathbf{x}}_{\mathrm{g}}}{\operatorname{argmin}}\left(C(\hat{\mathbf{x}}_{\mathrm{g}}, \mathbf{p}, \mathbf{v})\right), \tag{2}$$

where $\mathbf{p} = \left(\mathbf{p}_1^{\mathrm{T}}, \mathbf{p}_2^{\mathrm{T}}, \ldots, \mathbf{p}_N^{\mathrm{T}}\right)^{\mathrm{T}}$.

The selection of the appropriate cost functions is not trivial in the general case, because it largely depends on a priori data about the position of target point $\mathbf{x}_{\mathrm{g}}$, its projections $\mathbf{p}_i$, transformations P_i and the statistics of coordinate measurement errors. If we assume that the deviation of the measured image coordinates $\mathbf{p}_i$ from their true values $\bar{\mathbf{p}}_i$ follows the Gaussian distribution $N(\mathbf{0}, \mathbf{\Sigma}_{\mathbf{p}_i})$, then we can use the Maximum Likelihood Estimator (MLE), which minimizes the Mahalanobis distance in the image plane [21]

$$C = \sum_{i=1}^{N} \left\| \mathbf{p}_i - P_i \circ E_i\left(\hat{\mathbf{x}}_g\right) \right\|_{\Sigma}^{2} = \sum_{i=1}^{N} \left(\mathbf{p}_i - \hat{\mathbf{p}}_i\right)^{\mathrm{T}} \mathbf{\Sigma}_{\mathbf{p}_i}^{-1} \left(\mathbf{p}_i - \hat{\mathbf{p}}_i\right), \tag{3}$$

where $\hat{\mathbf{p}}_i = P_i \circ E_i\left(\hat{\mathbf{x}}_g\right)$ is the estimated position of the 3D point $\hat{\mathbf{x}}_g$ projection on the image plane of i-th camera, and $\mathbf{\Sigma}_{\mathbf{p}_i}^{-1}$ is the inverse (rank-constrained generalized inverse) covariation matrix of coordinate measurement error for $\mathbf{p}_i$, $\mathbf{\Sigma}_{\mathbf{p}} = \mathrm{diag}\left(\mathbf{\Sigma}_{\mathbf{p}_1}, \mathbf{\Sigma}_{\mathbf{p}_2}, \ldots, \mathbf{\Sigma}_{\mathbf{p}_N}\right)$. Given by nonlinear minimization of the cost function (3), the estimate $\hat{\mathbf{x}}_g$ is asymptotycally unbiased and effective in the limit of small noise. Furthermore, this algorighm allows to achieve the theoretical bound for minimal variance (the Cramer-Rao lower bound) [[18]], [[21]].

We can use the generic mathematical model described above to evaluate 3D coordinates uncertainty of individual points $\hat{\mathbf{x}}$ for different types of multi-view (or equivalent to multi-view) systems, including the system presented in the Figure 3. Error evaluation of complex characteristics measurement (distances, areas or volumes) for foreign inclusions or defects may be performed similarly based on the estimated uncertainties for individual points of the detected defect contour. It is important that the individual point errors are uncorrelated in this case. However, when measuring distance and area on the surface of known shape, it becomes possible to take shape information into consideration on the triangulation stage and significantly improve accuracy. This approach requires reconsidering the above mathematical model.

3. 3D point reconstruction on the surface of known shape

Consider the case of a parametric surface in 3D space: point $\mathbf{x}$ belongs to the surface if its coordinates are defined by a parametric equation $\mathbf{x} = S(\mathbf{a})$ of position parameters $\mathbf{a} = (a,b)^{\mathrm{T}}$. Note that point coordinates $\mathbf{x}$ are given in the surface CS, global coordinates may be calculated as $\mathbf{x}_g = E_s(\mathbf{x}) = \mathbf{R}_s\mathbf{x} + \mathbf{t}_s$, camera-centered coordinates – using relation (1). Next, combine the set of parameters for $\mathbf{R}_s$ and $\mathbf{t}_s$ together with additional parameters included in function $S(\mathbf{a})$ (such as radius of a cylindrical or spherical surface) and denote it as $\mathbf{v}_s$. Then we represent the triangulation algorithm in the following form

$$\hat{\mathbf{a}} = T(\mathbf{p}, \mathbf{v}, \mathbf{v}_s) = \underset{\hat{\mathbf{a}}}{\operatorname{argmin}}\left(C(\hat{\mathbf{a}}, \mathbf{p}, \mathbf{v}, \mathbf{v}_s)\right). \tag{4}$$

To calculate the 3D coordinates $\mathbf{x}$ from N registered images, as before, the best solution is to use an algorithm that minimizes the Mahalanobis distance (3) in the image planes, i.e.

$$C = \sum_{i=1}^{N} \left\| \mathbf{p}_i - P_i \circ E_i \circ E_s \circ S\left(\hat{\mathbf{a}}\right) \right\|_{\Sigma}^{2} = \sum_{i=1}^{N} \left(\mathbf{p}_i - \hat{\mathbf{p}}_i\right)^{\mathrm{T}} \mathbf{\Sigma}_{\mathbf{p}_i}^{-1} \left(\mathbf{p}_i - \hat{\mathbf{p}}_i\right). \tag{5}$$

In contrast to equations (2,3), the result of the minimization procedure according to the equations (4,5) is the estimate of parameters $\hat{\mathbf{a}}$, which defines the point position on the surface. Hence, we should use the known transformation $\hat{\mathbf{x}}_g = E_s\left(S\left(\hat{\mathbf{a}}\right)\right)$ to convert the estimated position parameters to the global coordinates .

4. Uncertainty evaluation method

The development of the uncertainty evaluation technique requires the analysis of the factors affecting the accuracy of 3D coordinates measurements for individual points. Main factors are:

- method error, caused by limitations of the considered geometric model of the image registration, simplifications and 3D reconstruction algorithm;
- random error of corresponding point positions on the image planes, which depends on the chosen correspondence technique (manual or automatic) and an image quality;
- random or systematic error of the parameters for the used mathematical model (calibration errors).

In the case of the improved triangulation algorithm (4), we should add the surface orientation and the parameters error to the factors described above. Since the analysis of the impact of the methodological error cannot be considered wthin the developed mathematical model, we further restrict our investigations to the second and third factors.

The estimation error of 3D coordinates $\hat{\mathbf{x}}$ calculated by the conventional triangulation algorithm (2) and the uncertainties of image point positions $\mathbf{p}_i$ are related via the backward propagation theorem [[18]], [[21]], [[22]]. One can differentiate the cost function with respect to $\mathbf{p}$ and perform some manipulations to derive the expression for the partial derivates of the bias and the error covariance matrix for the 3D coordinates. A similar approach can be applied to evaluate the error caused by deviations of calibration parameters $\mathbf{v}$ [[23]], [[24]].

If we consider the dependence of the estimation error of $\hat{\mathbf{x}}$ calculated by the triangulation algorithm (4) from deviations of $\mathbf{p}_i$, we should first solve this problem for the estimation error of position parameters $\hat{\mathbf{a}}$. Use the same theorem as before, and, after substitution the expression for the cost function (5) and algebraic transformations, we obtain the following expressions

$$\frac{\partial \hat{\mathbf{a}}}{\partial \mathbf{p}} = \left(\frac{\partial \mathbf{p}}{\partial \mathbf{a}}^{\mathrm{T}} \boldsymbol{\Sigma}_{\mathbf{p}}^{-1} \frac{\partial \mathbf{p}}{\partial \mathbf{a}} \right)^{-1} \frac{\partial \mathbf{p}}{\partial \mathbf{a}}^{\mathrm{T}} \boldsymbol{\Sigma}_{\mathbf{p}}^{-1}, \tag{6}$$

$$\boldsymbol{\Sigma}_{\mathbf{a}} = \frac{\partial \hat{\mathbf{a}}}{\partial \mathbf{p}} \boldsymbol{\Sigma}_{\mathbf{p}} \frac{\partial \hat{\mathbf{a}}}{\partial \mathbf{p}}^{\mathrm{T}} = \left(\frac{\partial \mathbf{p}}{\partial \mathbf{a}}^{\mathrm{T}} \boldsymbol{\Sigma}_{\mathbf{p}}^{-1} \frac{\partial \mathbf{p}}{\partial \mathbf{a}} \right)^{-1}, \tag{7}$$

where $\frac{\partial \mathbf{p}}{\partial \mathbf{a}} = \left(\frac{\partial \mathbf{p}_1}{\partial \mathbf{a}}^{\mathrm{T}}, \frac{\partial \mathbf{p}_2}{\partial \mathbf{a}}^{\mathrm{T}}, \ldots, \frac{\partial \mathbf{p}_N}{\partial \mathbf{a}}^{\mathrm{T}} \right)^{\mathrm{T}}$, $\frac{\partial \mathbf{p}_i}{\partial \mathbf{a}} = \frac{\partial P_i}{\partial \mathbf{x}_i} \frac{\partial E_i}{\partial \mathbf{x}_{\mathrm{g}}} \frac{\partial E_s}{\partial \mathbf{x}} \frac{\partial S}{\partial \mathbf{a}}$.

The posterior covariance matrix $\boldsymbol{\Sigma}_{\mathbf{a}}$ for the estimate of position parameters $\hat{\mathbf{a}}$ in the equation (7) is given for the case where system calibration performed without errors and transformations P_i, E_i determined exactly. Furthermore, transformations E_s and S are asssumed to be known without uncertainties too.

In order to derive the equation for the cost function (5) considering calibration errors and uncertainties of surface orientation, let the mapping $(P_i \circ E_i)$ be differentiable with respect to $\mathbf{v}$ in a neighborhood of the true value $\bar{\mathbf{v}}$ and let the mapping $(E_s \circ S)$ be differentiable with respect to $\mathbf{v}_{\mathrm{s}}$ in the neighborhood of $\bar{\mathbf{v}}_{\mathrm{s}}$. Then one can use the first-order approximation to write

$$C = \sum_{i=1}^{N} \left\| \mathbf{p}_i - P_i \circ E_i \left(E_s \circ S(\hat{\mathbf{a}}) - \frac{\partial (E_s \circ S)}{\partial \mathbf{v}_{\mathrm{s}}} (\mathbf{v}_{\mathrm{s}} - \bar{\mathbf{v}}_{\mathrm{s}}) \right) - \frac{\partial (P_i \circ E_i)}{\partial \mathbf{v}} (\mathbf{v} - \bar{\mathbf{v}}) \right\|_{\Sigma}^{2}. \tag{8}$$

After substitution and some manipulations we obtain the expression for the partial derivates of the bias and the covariance matrix

$$\frac{\partial \hat{\mathbf{a}}}{\partial \mathbf{v}} = -\mathbf{N}^{-1} \frac{\partial \mathbf{p}}{\partial \mathbf{a}}^{\mathrm{T}} \boldsymbol{\Sigma}_{\mathrm{p}}^{-1} \frac{\partial \mathbf{p}}{\partial \mathbf{v}}, \quad \frac{\partial \hat{\mathbf{a}}}{\partial \mathbf{v}_{\mathrm{s}}} = -\mathbf{N}^{-1} \frac{\partial \mathbf{p}}{\partial \mathbf{a}}^{\mathrm{T}} \boldsymbol{\Sigma}_{\mathrm{p}}^{-1} \frac{\partial \mathbf{p}}{\partial \mathbf{v}_{\mathrm{s}}}, \tag{9}$$

$$\boldsymbol{\Sigma}_{\mathrm{a}} = \frac{\partial \hat{\mathbf{a}}}{\partial \mathbf{p}} \boldsymbol{\Sigma}_{\mathrm{p}} \frac{\partial \hat{\mathbf{a}}}{\partial \mathbf{p}}^{\mathrm{T}} + \frac{\partial \hat{\mathbf{a}}}{\partial \mathbf{v}} \boldsymbol{\Sigma}_{\mathrm{v}} \frac{\partial \hat{\mathbf{a}}}{\partial \mathbf{v}}^{\mathrm{T}} + \frac{\partial \hat{\mathbf{a}}}{\partial \mathbf{v}_{\mathrm{s}}} \boldsymbol{\Sigma}_{\mathrm{v_s}} \frac{\partial \hat{\mathbf{a}}}{\partial \mathbf{v}_{\mathrm{s}}}^{\mathrm{T}}, \tag{10}$$

where $\frac{\partial \mathbf{p}_i}{\partial \mathbf{v}} = \frac{\partial P_i}{\partial \mathbf{x}_i} \frac{\partial E_i}{\partial \mathbf{v}} + \frac{\partial P_i}{\partial \mathbf{v}}$, $\frac{\partial \mathbf{p}_i}{\partial \mathbf{v}_{\mathrm{s}}} = \frac{\partial P_i}{\partial \mathbf{x}_i} \frac{\partial E_i}{\partial \mathbf{x}_{\mathrm{g}}} \left(\frac{\partial E_s}{\partial \mathbf{x}} \frac{\partial S}{\partial \mathbf{v}_{\mathrm{s}}} + \frac{\partial E_s}{\partial \mathbf{v}_{\mathrm{s}}} \right)$, $\mathbf{N}$; $\frac{\partial \mathbf{p}}{\partial \mathbf{v}}$ and $\frac{\partial \mathbf{p}}{\partial \mathbf{v}_{\mathrm{s}}}$ are combinations of $\frac{\partial \mathbf{p}_i}{\partial \mathbf{v}}$ and $\frac{\partial \mathbf{p}_i}{\partial \mathbf{v}_{\mathrm{s}}}$ similarly to $\frac{\partial \mathbf{p}}{\partial \mathbf{a}}$ above.

Thus, the combination of the expression (10) and partial derivatives (6, 9) produces the final formula for the posterior covariation matrix $\boldsymbol{\Sigma}_{\mathrm{a}}$ based on the main factors affecting measurement accuracy. To get a similar expression for the posterior covariation matrix $\boldsymbol{\Sigma}_{\mathrm{x_g}}$ for the coordinates in the GCS, we differentiate $\hat{\mathbf{x}}_{\mathrm{g}} = E_s\left(S\left(\hat{\mathbf{a}}\right)\right)$

$$\frac{\partial \hat{\mathbf{x}}_{\mathrm{g}}}{\partial \mathbf{p}} = \frac{\partial \mathbf{x}_{\mathrm{g}}}{\partial \mathbf{a}} \frac{\partial \hat{\mathbf{a}}}{\partial \mathbf{p}}, \quad \frac{\partial \hat{\mathbf{x}}_{\mathrm{g}}}{\partial \mathbf{v}} = \frac{\partial \mathbf{x}_{\mathrm{g}}}{\partial \mathbf{a}} \frac{\partial \hat{\mathbf{a}}}{\partial \mathbf{v}}, \quad \frac{\partial \hat{\mathbf{x}}_{\mathrm{g}}}{\partial \mathbf{v}_{\mathrm{s}}} = \frac{\partial \mathbf{x}_{\mathrm{g}}}{\partial \mathbf{a}} \frac{\partial \hat{\mathbf{a}}}{\partial \mathbf{v}_{\mathrm{s}}} + \frac{\partial \mathbf{x}_{\mathrm{g}}}{\partial \mathbf{v}_{\mathrm{s}}}, \tag{11}$$

$$\boldsymbol{\Sigma}_{\mathrm{x_g}} = \frac{\partial \hat{\mathbf{x}}_{\mathrm{g}}}{\partial \mathbf{p}} \boldsymbol{\Sigma}_{\mathrm{p}} \frac{\partial \hat{\mathbf{x}}_{\mathrm{g}}}{\partial \mathbf{p}}^{\mathrm{T}} + \frac{\partial \hat{\mathbf{x}}_{\mathrm{g}}}{\partial \mathbf{v}} \boldsymbol{\Sigma}_{\mathrm{v}} \frac{\partial \hat{\mathbf{x}}_{\mathrm{g}}}{\partial \mathbf{v}}^{\mathrm{T}} + \frac{\partial \hat{\mathbf{x}}_{\mathrm{g}}}{\partial \mathbf{v}_{\mathrm{s}}} \boldsymbol{\Sigma}_{\mathrm{v_s}} \frac{\partial \hat{\mathbf{x}}_{\mathrm{g}}}{\partial \mathbf{v}_{\mathrm{s}}}^{\mathrm{T}}, \tag{12}$$

where $\frac{\partial \mathbf{x}_{\mathrm{g}}}{\partial \mathbf{a}} = \frac{\partial E_s}{\partial \mathbf{x}} \frac{\partial S}{\partial \mathbf{a}}$, $\frac{\partial \mathbf{x}_{\mathrm{g}}}{\partial \mathbf{v}_{\mathrm{s}}} = \frac{\partial E_s}{\partial \mathbf{x}} \frac{\partial S}{\partial \mathbf{v}_{\mathrm{s}}} + \frac{\partial E_s}{\partial \mathbf{v}_{\mathrm{s}}}$.

Moreover, a number of additional factors also reduces the measurement accuracy for distances (the error of calibration parameters leads to the correlation of errors of the individual points 3D coordinates which must be taken into account when calculating the distance error), arc lengths (the sum-of-segments approximation) and areas (the sum-of-triangles approximation). As shown in [[14]], the area measurement error caused by the sum-of-triangles approximation on cylindrical and spherical surfaces is less than 1%, and is not the main contributor.

Thus, the error evaluation of complex characteristics measurement for foreign inclusions or defects requires calculation of joint partial derivatives and the joint covariance matrix for all points of the investigated object contour according to equations (6, 9–12). As an example, consider the solution of this problem for the area of triangle defined by three points $\mathbf{x}_{\mathrm{g}}^1$, $\mathbf{x}_{\mathrm{g}}^2$ and $\mathbf{x}_{\mathrm{g}}^3$ on the surface of interest. The area of this triangle can be calculated as

$$S = \left\| \left(\mathbf{x}_{\mathrm{g}}^2 - \mathbf{x}_{\mathrm{g}}^1\right) \times \left(\mathbf{x}_{\mathrm{g}}^3 - \mathbf{x}_{\mathrm{g}}^1\right) \right\| / 2. \tag{13}$$

We can also obtain partial derivatives $\frac{\partial S}{\partial \mathbf{x}_{\mathrm{g}}^j}$, $j = 1,2,3$ from formula (13). In order to evaluate the joint covariance matrix $\boldsymbol{\Sigma}_{\mathrm{x_g^{1,2,3}}}$ for $\mathbf{x}_{\mathrm{g}}^1$, $\mathbf{x}_{\mathrm{g}}^2$ and $\mathbf{x}_{\mathrm{g}}^3$, some modifications of the equation (12) should be done. In particular, replace $\frac{\partial \hat{\mathbf{x}}_{\mathrm{g}}}{\partial \mathbf{p}}$ with $\mathrm{diag}\left(\frac{\partial \hat{\mathbf{x}}_{\mathrm{g}}^1}{\partial \mathbf{p}^1}, \frac{\partial \hat{\mathbf{x}}_{\mathrm{g}}^2}{\partial \mathbf{p}^2}, \frac{\partial \hat{\mathbf{x}}_{\mathrm{g}}^3}{\partial \mathbf{p}^3} \right)$, $\boldsymbol{\Sigma}_{\mathrm{p}}$ with $\mathrm{diag}\left(\boldsymbol{\Sigma}_{\mathrm{p}^1}, \boldsymbol{\Sigma}_{\mathrm{p}^2}, \boldsymbol{\Sigma}_{\mathrm{p}^3} \right)$, $\frac{\partial \hat{\mathbf{x}}_{\mathrm{g}}}{\partial \mathbf{v}}$

with $\left(\frac{\partial\hat{\mathbf{x}}_g^1}{\partial\mathbf{v}},\frac{\partial\hat{\mathbf{x}}_g^2}{\partial\mathbf{v}},\frac{\partial\hat{\mathbf{x}}_g^3}{\partial\mathbf{v}}\right)^T$ and $\frac{\partial\hat{\mathbf{x}}_g}{\partial\mathbf{v}_s}$ with $\left(\frac{\partial\hat{\mathbf{x}}_g^1}{\partial\mathbf{v}_s},\frac{\partial\hat{\mathbf{x}}_g^2}{\partial\mathbf{v}_s},\frac{\partial\hat{\mathbf{x}}_g^3}{\partial\mathbf{v}_s}\right)^T$. Next, write the expression for the posterior variance $\sigma_S{}^2$ of the area measurement error

$$\sigma_S{}^2=\left(\frac{\partial S}{\partial\mathbf{x}_g^1}\quad\frac{\partial S}{\partial\mathbf{x}_g^2}\quad\frac{\partial S}{\partial\mathbf{x}_g^3}\right)\mathbf{\Sigma}_{\mathbf{x}_g^{1,2,3}}\left(\frac{\partial S}{\partial\mathbf{x}_g^1}\quad\frac{\partial S}{\partial\mathbf{x}_g^2}\quad\frac{\partial S}{\partial\mathbf{x}_g^3}\right)^T. \tag{14}$$

One can derive the equation for the measurement error of the area of a complex figure, which is calculated as the sum of the areas of individual triangles, similar to (14), but this is more complicated because of the presence of joint vertices of the neighboring triangles.

5. Experiments and numerical modelling

To demonstrate the proposed method we apply it for the accuracy evaluation of the point positions and the area measurements using the single-imager stereoscopic endoscope with the following parameters: CCD sensor 1/10" (752×576 elements), field of view 60° of each channel, base distance 3 mm. The probe axis coincides with the axis of the test object presented by the graph paper tube with a radius of 20 mm (see Figure 5). Analyzed points are evenly spaced on the surface of the cylinder with the step 5 mm. Thus, the area of each triangle (half of the square shown in Figure 5) is equal to 12.5 mm^2.

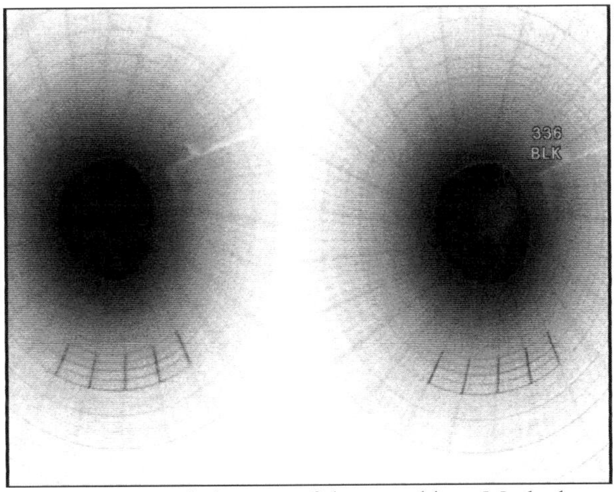

Figure 5. Stereoscopic images of the test object. Marked squares.

We calculate the error variance of the triangles areas caused by the random uncorrelated error of the corresponding point positions with the variance $\sigma_p{}^2=0.5$ for both coordinates u,v and both images. The results are presented for the conventional triangulation algorithm (2) (in the Figure 6) and for the surface-adopted triangulation algorithm (4) (in the Figure 7).

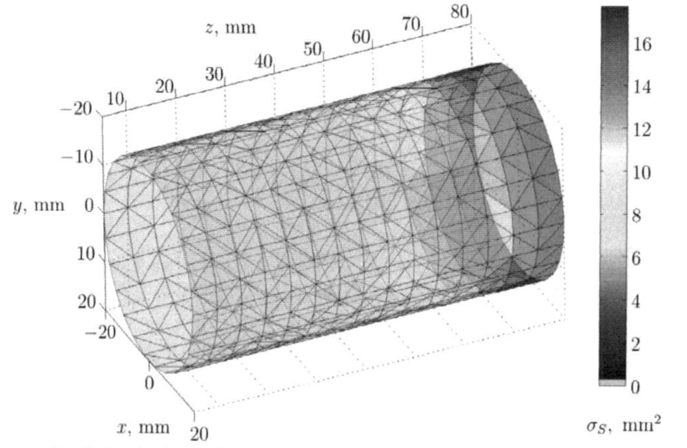

Figure 6. The standard deviation of the area measurement for the conventional triangulation algorithm. The true value of the each triangle area is equal to 12.5 mm^2.

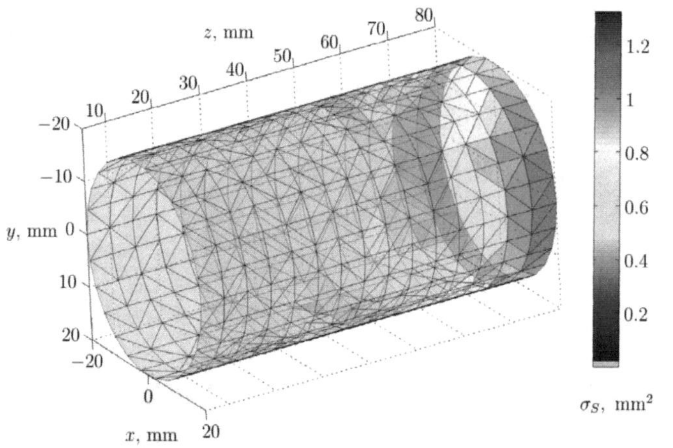

Figure 7. The standard deviation of the area measurement for the surface-adopted triangulation algorithm. The true value of the each triangle area is equal to 12.5 mm^2.

The color of each triangle represents the standard deviation σ_S of the area measurement according to the color bar on the right side of every figure. The triangles shown in gray color are out of view for at least one of the channels. The origin of the CS in the figures is the center of probe located between two camera centers. Note that taking shape information into consideration on the triangulation stage radically improves measurement accuracy (from $\sigma_S = 8.3\ \text{mm}^2$ down to $\sigma_S = 0.5\ \text{mm}^2$ for the triangle placed at a distance of 50 mm from the origin).

Next, we use the developed mathematical models to examine the influence of errors of the object surface orientation relative to the probe. Below are the results of the error estimation for the 3D coordinates of the individual points and the areas of triangles caused by the random rotation with a variance $\sigma_\alpha{}^2$ for both angles α_x and α_y (see Figure 8) and the random decentering with a variance $\sigma_{\Delta c}{}^2$ for both coordinates x, y (see Figure 9).

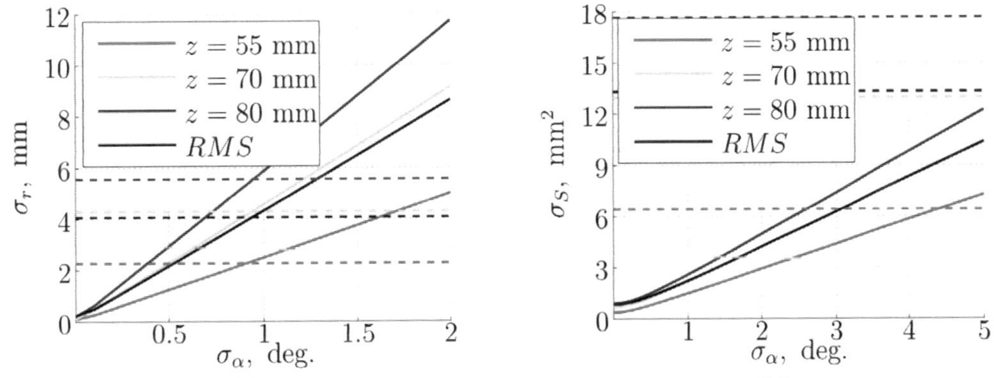

(a) (b)

Figure 8. The square root of the largest eigenvalues σ_r of the covariation matrices $\mathbf{\Sigma}_{\mathbf{x}_g}$ for the point positions (a) and the standard deviation σ_S of triangles' area (b) calculated for the different values of the standard deviation σ_α of the rotation angles α_x and α_y. Solid lines correspond to the surface-adopted triangulation algorithm, dotted lines correspond to the conventional one.

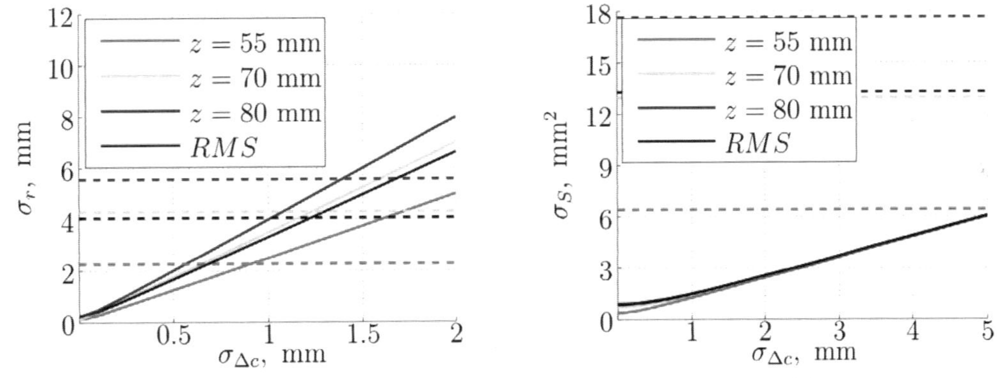

(a) (b)

Figure 9. The square root of the largest eigenvalues σ_r of the covariation matrices $\mathbf{\Sigma}_{\mathbf{x}_g}$ for the point positions (a) and the standard deviation σ_S of triangles' area (b) calculated for the different values of the standard deviation $\sigma_{\Delta c}$ of decentering for both coordinates x, y. Solid lines correspond to the surface-adopted triangulation algorithm, dotted lines correspond to the conventional one.

Figures 8a and 9a show the square root of the largest eigenvalues σ_r of the covariation matrices $\mathbf{\Sigma}_{\mathbf{x}_g}$ for the points at different distances from the origin (for example, $z = 55$ mm) and the RMS value (denoted as 'RMS' in the figure legend) of σ_r for all visible points on the cylindrical surfaces with *z*-coordianate in the range 10-80 mm. Hence, the values of σ_r correspond to the major axis of the error ellipsoid, i.e. the direction along which the 3D point position has the maximum variance [[25]]. One can see, that in **Upper Saddle River** creasing uncertainty of the surface orientation leads to the linear growth of the errors of 3D point positions reconstructed using the surface-adapted triangulation algorithm. Thus, it makes possible to find the lagrest tolerable deviation of the surface, whereby this algorithm still outperforms the conventional one (for example, find the value of σ_α in which a solid line crosses a dotted line in the Figure 8a).

Figures 8b and 9b show the standard deviation σ_S of the triangles' area at the same distances. Note that the errors of the area measurement increase much slower than in the previous case due to the strong correlation of the individiual point position errors. Consequently, the surface-adapted triangulation algorithm is preferrable to the conventional one, even with the large uncertatinties of surface orientation relative to the probe. This is especially evident for decentering of the imager (see Figure 9b).

References

[1] Chang C I 2003 *Hyperspectral imaging: techniques for spectral detection and classification* (New York: Springer)
[2] Shepherd G 2002 *Spectral imaging of the atmosphere* (New York: Academic press)
[3] Denes L J, Gottlieb M S, Kaminsky B 1998 *Opt. Eng.* **37** 1262-67
[4] Goutzulis A, Rape D 1994 *Design and fabrication of acousto-optic devices* (New York: Dekker)
[5] Pozhar V E, Machihin A S 2012 *Appl. Opt.* **51**(19) 4513-19
[6] Pannel C N, Wachman E S, Farkas D L, Ward J, Seale W 2005 *Proc. SPIE* **6088** 165-69
[7] Pozhar V E, Machihin A S, Batshev V I 2013 *Photonics and Lasers in Medicine* **2**(2) 153–57
[8] Gupta N 2008 *Proc. SPIE* **6940** 1-10
[9] Pozhar V E, Machihin A S 2012 *AIP Conf. Proc.* **1433** 65-67
[10] Yoon S, Thai C 2009 *Proc. ASABE Annual Int. Meeting* **096583** 1-12
[11] Viskovatykh A V, Machikhin A S, Pozhar V E, Pustovoit V I, Viskovatykh D A 2014 *Tech. Phys. Lett.* **40**(2) 157-60
[12] Ezhov V A, Studentsov S A 2005 *Proc. SPIE.* **5821** 102-16
[13] Machikhin A S, Pozhar V E 2014 *Tech. Phys. Lett.* **40**(18) to be published
[14] Gorevoy A V, Machikhin A S and Perfilov A M 2014 *Sci. and Tech. J. of Inf. Technology, Mechanics and Optics* **4**(92) 140–48
[15] Kannala J, Heikkilä J and Brandt S S 2009 *Geometric camera calibration Wiley Encyclopedia of Computer Science and Engineering* ed. B W Wah (Hoboken: Wiley)
[16] Ramalingam S 2006 *Generic Imaging Models: Calibration and 3D Reconstruction Algorithms* (Grenoble: Institut National Polytechnique de Grenoble)
[17] Kannala J and Brandt S S 2006 *IEEE Trans. Pattern Anal. and Machine Intelligence* **28**(8) 1335–40
[18] Hartley R I and Zisserman 2000 *A Multiple View Geometry* 2nd ed. (Cambridge: Cambridge University Press)
[19] Forsyth D A and Ponce J 2012 *Computer Vision: a Modern Approach* 2nd ed (Upper Saddle River: Prentice-Hall)
[20] Zhang Z 2000 *IEEE Trans. on Pattern Anal. and Machine Intelligence* **22**(11) 1330–34
[21] Kanatani K 2005 *Statistical Optimization for Geometric Computation: Theory and Practice* (Mineola: Dover Publications)
[22] Zhang Z 1998 *Int. J. of Comp. Vis.* **27**(2) 161–95
[23] Abraham S and Förstner W 1998 *Proc. DAGM Symp. Mustererkennung* pp 117–24
[24] Gorevoy A V and Koluchkin V Ya 2013 *Eng. J.: Science and Innovation* **9**(21) 45
[25] Johnson R A and Wichern D W 2007 *Applied Multivariate Statistical Analysis* 6th ed. (Upper Saddle River: Pearson Prentice Hall)

An impact of multiple wave reflections in a flat sample on material parameter reconstruction using THz pulsed spectroscopy

Kirill I Zaytsev[1], Konstantin P Tsapenko[2], Pavel A Nosov and Stanislav O Yurchenko[3]

Bauman Moscow State Technical University, 105005, 2nd Baumanskaya str. 5, Moscow, Russia

E-mail: [1]kirzay@gmail.com, [2]kostyatsapenko@gmail.com, [3]st.yurchenko@mail.ru

Abstract. The present paper is dedicated to studying the accuracy of sample material parameter reconstruction using terahertz (THz) pulsed spectroscopy. The technique for characterizing the material parameters of thin flat samples allowing to take into account multiple wave reflections in a flat have been considered. While transmitting through the resonant sample, THz pulse undergoes multiple reflections, which result in satellite pulses in THz waveform. The accuracy of material parameter reconstruction strongly depends on the number of satellite pulses. We have analytically estimated an impact of satellite pulses on material parameter reconstruction by considering the local linearization of theoretical sample transfer function based on the model of quasi-Fabry-Perot-resonator.

1. Introduction

Terahertz (THz) pulsed spectroscopy (TPS) appeared in 1975 as a result of Auston's research of semiconductor photoconductivity under the femtosecond optical pumping [1]. Since the emergence of highly stable portable femtosecond lasers TPS becomes a powerful tool for fundamental and applied research. TPS provides an ability both for sample material parameter reconstruction [2–12] and for studying the internal structure of the sample (T-Ray tomography) [13–15].

Regardless the type of data we are measuring the main principle of TPS operation could be formulated as follows. THz pulse is generated with one of the existing techniques [1,16–19], and the electric field of THz pulse is detected [20] with a high time resolution after the transmission of pulse through the sample or after the reflection of pulse from its surface. By applying a fast Fourier transform to time-domain data, it is possible to analyze the complex transmission and reflection coefficients of the sample [12], in a wide frequency range, as well as to study the sample pulse response [13,14]. All the specified characteristics form an initial conditions for TPS inverse problem solutions. There are many applications for TPS systems, including condensed matter spectroscopy [21], gas sensing [22], spectroscopy of colloid crystals [23], protein crystals [24–26], medical diagnosis [27–36], security task [37–39] etc. Recently, TPS has become a useful tool for nondestructive evaluation of polymer composite constructions [40–42].

Owing to the wide variety of TPS applications, many TPS signal processing problems appear, for instance, the inverse scattering problems. Reconstructing the resonant sample material

Content from this work may be used under the terms of the Creative Commons Attribution 3.0 licence. Any further distribution of this work must maintain attribution to the author(s) and the title of the work, journal citation and DOI.
Published under licence by IOP Publishing Ltd

parameters in transmission mode is a representative example of TPS inverse problem [2–12]. Most TPS inverse problems, regardless the specificity of reconstructed data, are ill-posed, and the accuracy of reconstruction depends on a large number of factors [11,12]. The stability of the TPS inverse problem solution in presence of different factors must be systematically studied for correct implementation of TPS signal processing.

In this paper, we present new results for sample material parameter reconstruction using TPS. We are analyzing a commonly used algorithm for resonant sample material parameter reconstruction based on quasi-Fabry-Perot model, which takes into account THz pulse transmission through the resonance sample [7,10,12]. While transmitting through the resonant sample THz pulse undergoes multiple reflections, thus, the THz waveform contain finite number of satellite THz pulses. Depending on the number of these satellite pulses we could obtain different reconstruction accuracy. We are estimating the impact of satellite pulse number on the accuracy of material parameter reconstruction utilizing the local linearization of quasi-Fabry-Perot model.

2. Quasi-Fabry-Perot approach for material parameter reconstruction

At first, let us describe a common material parameter reconstruction procedure [7,10,12], which we are considering in the present paper. This technique helps to reconstruct the complex refractive index of the flat homogeneous sample

$$\widetilde{n}(\nu) = n'(\nu) - in''(\nu) = n'(\nu) - i\frac{\alpha(\nu)\, c_0}{2\pi\nu}, \tag{1}$$

where ν is a frequency of electromagnetic wave, c_0 is a speed of light in vacuum, $n'(\nu)$ and $n''(\nu)$ are real and imaginary parts of complex refractive index, and $\alpha(\nu)$ is an absorption coefficient. The technique we are describing is appropriate for studying the resonant flat samples, transparent in THz frequency range and having the thickness

$$l > l_{min} = \frac{c_0}{\nu_{max} n_{min}}, \tag{2}$$

where n_{min} is the minimal value of sample refractive index within the range of system sensitivity, $[\nu_{min}, \nu_{max}]$.

Reconstruction of sample material parameters, $\widetilde{n}_1$, is produced by minimization of the following error functional

$$\widetilde{n}_1(\nu) = \mathrm{argmin}_{\widetilde{n}_1(\nu)} \left[M(\nu, \widetilde{n}_1) + A(\nu, \widetilde{n}_1) \right], \tag{3}$$

where

$$M(\nu, \widetilde{n}_1) = \left| \left| \widetilde{H}_{exp}(\nu) \right| - \left| \widetilde{H}_{th}(\nu, \widetilde{n}_1) \right| \right|^2, \qquad A(\nu, \widetilde{n}_1) = \left| \phi\left[\widetilde{H}_{exp}(\nu) \right] - \phi\left[\widetilde{H}_{th}(\nu, \widetilde{n}_1) \right] \right|^2, \tag{4}$$

$\widetilde{H}_{exp}(\nu)$ and $\widetilde{H}_{th}(\nu, \widetilde{n}_1)$ correspond to experimental and theoretical transfer functions of the sample, $|...|$ and $\phi[...]$ represent estimation of transfer function modulus and phase.

Functional minimization (equation (3)) could be implemented separately for different Fourier-domain frequencies ν. The details of functional minimization process are described in paper [12]. The experimental complex transfer function $\widetilde{H}_{exp}(\nu)$ can be calculated using the relation

$$\widetilde{H}_{exp}(\nu) = \frac{\widetilde{E}_s(\nu)}{\widetilde{E}_r(\nu)}, \tag{5}$$

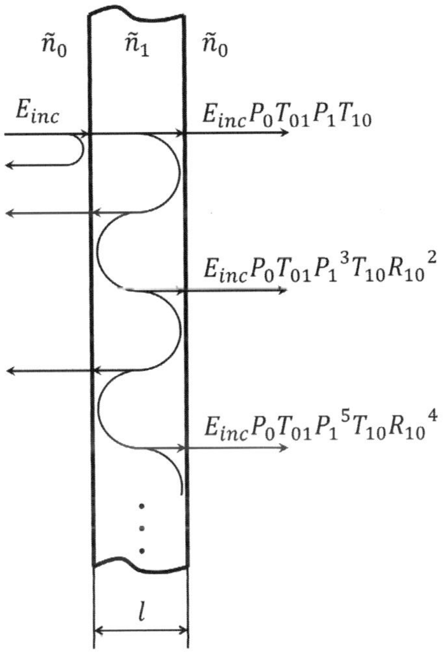

Figure 1. Schematic representation of multiple light scattering in the flat.

where tildes stand for Fourier transform, $\widetilde{E}_r(\nu) = F_t^{-1}[E_r(t)]$ and $\widetilde{E}_s(\nu) = F_t^{-1}[E_s(t)]$, of reference and sample waveforms, $E_r(t)$ and $E_s(t)$. Theoretical transfer function is based on the quasi-Fabry-Perot model [12] of the wave propagation through the sample (figure 1):

$$\widetilde{H}_{th}(\nu, \widetilde{n}_1) = \frac{4\widetilde{n}_0\widetilde{n}_1}{(\widetilde{n}_0 + \widetilde{n}_1)^2} exp\left(-ik\left(\widetilde{n}_1 - \widetilde{n}_0\right) l\right) \left[1 + \sum_{\xi=1}^{k} \left(\frac{\widetilde{n}_0 - \widetilde{n}_1}{\widetilde{n}_0 + \widetilde{n}_1} exp\left(-ik\widetilde{n}_1 l\right)\right)^{2\xi}\right], \quad (6)$$

where k represents the number of satellite pulses in sample waveform, $E_s(t)$, l is a thickness of sample, and $\widetilde{n}_0$ is a refractive index of air. This equation is based on the Bouguer-Lambert law, describing wave propagation through dissipative medial, as well as on the Fresnel equations for light reflection and transmission through the interface of two media (we are utilizing Fresnel formulas for the case of normal light incidence on the surface of interface).

The sample waveform contains a limited number of satellite pulses, k, owing to the finiteness of the TPS time-domain window [12]. Therefore, a finite number of pulses should be taken into account during the $\widetilde{n}_1(\nu_t)$ computation. k can be easy calculated during the TPS waveform pre-processing $E_s(t)$ [7,10,12]. Obviously, the accuracy of material parameter reconstruction strongly depends on k. We are estimating this dependence by means of local linearization of sample transfer function model (6).

3. Local linearization of theoretical transfer function

Many factors impact the material parameter reconstruction accuracy, including the noises in TPS waveforms and fluctuations of various TPS parameters. Studying these factors is very important for correct implementation of TPS measurements [12]. All factors lead to appearance

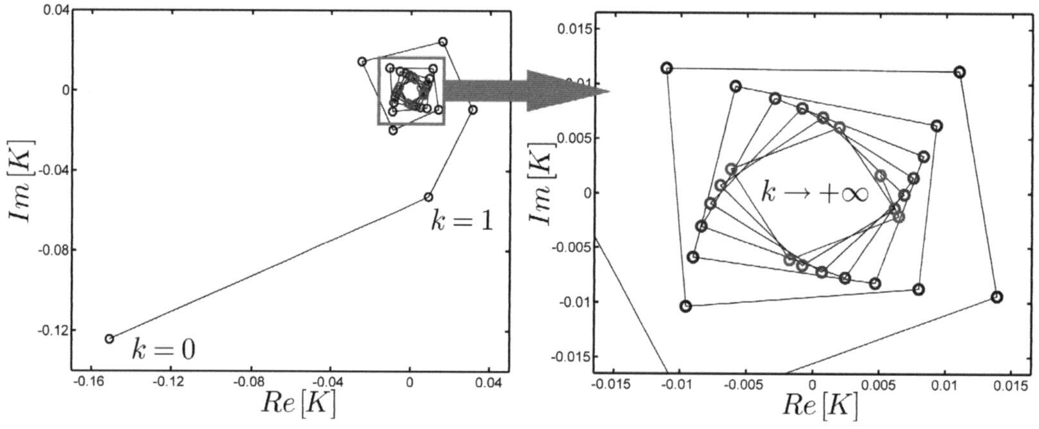

Figure 2. Local linearization of quasi-Fabry-Perot model: **(a)** shows K for all possible numbers of multiple resonances within the sample, $k = 0, 1, 2, ..., +\infty$, and **(b)** shown magnified part of complex plane centered near the zero-point and marked with red box in (a) panel.

of an error in TPS waveforms, ΔE, and, as a consequence, to appearance of an error in transfer function, $\Delta\widetilde{H}$. The latter one usually varies with frequency variations, $\Delta\widetilde{H} = \Delta\widetilde{H}(\nu)$, and it leads to appearance of an error in reconstructed material parameters, $\Delta\widetilde{n}_1 = \Delta\widetilde{n}_1(\nu)$.

We are estimating an impact of transfer function error, $\Delta\widetilde{H}$, on material parameter reconstruction error, $\Delta\widetilde{n}_1$, for different number of satellite pulses, k. Rigorous relation between $\Delta\widetilde{n}_1$ and $\Delta\widetilde{H}$ is defined by quasi-Fabry-Perot model (6), and it has a complex character in case $k > 0$. Therefore, we perform local linearization of (6)

$$\Delta\widetilde{n}_1(\nu) = K\Delta\widetilde{H}(\nu), \tag{7}$$

where K is a coefficient of local linearization depending on the condition of experiment, $K = K(\nu, k, l, \widetilde{n}_0, \widetilde{n}_1)$. We have calculated K for a certain experimental conditions ($\widetilde{n}_0 = 1.0$, $\widetilde{n}_1 = 2.00 + i0.477$, $\nu = 1.0$ THz, and $l = 0.1$ mm) and for various number of satellite pulses, k, in a sample signal, thus, the dependence $K(k)$ have been obtained. Figure 2 shows the results of $K(k)$ calculations.

With an increase of k the modulus of local linearization coefficients, $|K(k)|$, decays. Thus, the more satellite pulses we are detecting and taking into account during material parameter reconstruction, the higher reconstruction accuracy we could achieve for a certain power of noises in TPS waveforms, $E_r(t)$ and $E_s(t)$. Moreover, $K(k)$ dependence oscillates around zero point, $(0,0)$, of complex plane with an increase of k, therefore, the relation between the impact of transfer function error on real and imaginary parts, n' and n'', of material parameters differs for various number of satellite pulses. Simplest techniques for material parameter reconstruction allow reconstructing $\widetilde{n}_1$ using only main pulse of the sample waveform, while satellite pulses are being canceled by means of window filtering. Obviously, neglecting the satellite pulses could significantly reduce the computation complexity. At the same time it could strongly reduce the accuracy of material parameter reconstruction, as follows from figure 2.

4. Conclusion

We have studied the technique for material parameter reconstruction using TPS allowing to take into account multiple pulse reflection in a flat homogeneous sample. We have estimated an impact of the satellite pulses on the accuracy of material parameter reconstruction by applying the local linearization to the sample model which has the form of quasi-Fabry-Perot resonator. Strong dependence of the material parameter error on the number of satellite pulses has been observed. We have demonstrated that the average error of material parameter reconstruction reduces with an increase of satellite pulses number, thus, the more pulses we could detect with particular time-domain window of TPS, the higher accuracy of material parameter reconstruction we could achieve.

Acknowledgments

This work was supported by the grants of Russian Scientific Foundation, Project # 14-29-00277.

References

[1] Auston D H 1975 *Appl. Phys. Lett.* **26** 101–03
[2] Duvillaret L, Garet F, Coutaz J-L 1996 *IEEE J. Selected Topics in Quantum Electronics* **2** 739–46
[3] Duvillaret L, Garet F, Coutaz J-L 1999 *Appl. Opt.* **38** 409–15
[4] Dorney T D, Baraniuk R G, Mittleman D M 2001 *J. Opt. Soc. Am. A* **18** 1562–71
[5] Mickana S P, Lee K-S, Lu T M, Barnat E, Munch J, Abbott D, Zhang X-C 2001 *Proc. SPIE* **4591** 197–209
[6] Withayachumnankul W, Ferguson B, Rainsford T, Mickan S P, Abbott D 2005 *Proc. SPIE* **5840** 221–31
[7] Pupeza I, Wilk R, Koch M 2007 *Opt. Exp.* **15** 4335–50
[8] Balakrishnan J, Fischer B M, Abbott D 2010 *Opt. Commun.* **283** 2301–7
[9] Scheller M 2011 *Opt. Exp.* **19** 10647–55
[10] Zaytsev K I, Gavdush A A, Lebedev S P, Yurchenko S O 2014 *J. Phys. Conf. Ser.* **486** 012018
[11] Kruger M, Funkner S, Brundermann E, Havenith M 2011 *J. Infrared, Millimeter, and Terahertz Waves* **32** 699–715
[12] Zaytsev K I, Gavdush A A, Karasik V E, Alekhnovich V I, Nosov P A, Lazarev V A, Reshetov I V, Yurchenko S O 2014 *J. Appl. Phys.* **115** 193105
[13] Mittleman D M, Hunsche S, Boivin L, Nuss M C 1997 *Opt. Lett.* **22** 904–6
[14] Zaytsev K I, Karasik V E, Fokina I N, Alekhnovich V I 2013 *Opt. Eng.* **52** 068203
[15] Zaytsev K I, Chernomyrdin N V, Alekhnovich V I 2014 *J. Phys. Conf. Ser.* **486** 012010
[16] Auston D H, Smith P R 1983 *Appl. Phys. Lett.* **43** 631–33
[17] Klatt G, Hilser F, Qiao W, Beck M, Gebs R, Bartels A, Huska K, Lemmer U, Bastian G, Johnston M B, Fischer M, Faist J, Dekorsy T 2010 *Opt. Exp.* **18** 4939–47
[18] Hwang J S, Lin H-C, Chang C-K, Wang T S, Chang L-S, Chyi J-I, Liu W-S, Chen S-H, Lin H-H, Liu P-W 2007 *Opt. Exp.* **15** 5120–25
[19] Xu L, Zhang X-C, Auston, D H 1992 *Appl. Phys. Lett.* **61** 1992
[20] Lee Y-S 2009 *Principles of Terahertz Science and Technology* (New York: Springer)
[21] Auston D H, McAfee S, Shank C V, Ippen E P, Teschke O N 1978 *Solid State Electronics* **21** 147–50
[22] Mittleman D M, Jacobsen R H, Neelamani R, Baraniuk R G, Nuss M C 1998 *Appl. Phys. B* **67** 379–90
[23] Mittleman D M, Bertone J F, Jiang P, Hwang K S, Colvin V L 1999 *J. Chem. Phys.* **111** 345–54
[24] Lipps F, Levy S, Markelz A G 2012 *Phys. Chem.* **14** 6375–81
[25] He Y, Ku P I, Knab J R, Chen J Y, Markelz A G 2008 *Phys. Rev. Lett.* **101** 178103
[26] Markelz A G, Knab J R, Chen J Y, He Y *Chem. Phys. Lett.* **442** 413–17
[27] Woodward R M, Wallace V P, Pye R J, Cole B E, Arnone D D, Lineld E H, Pepper M 2003 *J. Investigative Dermatology* **120** 72–78
[28] Wallace V P, Woodward R M, Fitzgerald A J, Pickwell E, Pye R J, Arnone D A 2003 *Proc. SPIE* **4949** 353–59
[29] Wallace V P, Fitzgerald A J, Shankar S, Flanagan N, Pye R, Cluff J, Arnone D D 2004 *British Journal of Dermatology* **151** 424–32
[30] Zaytsev K I, Kudrin K G, Koroleva S A, Fokina I N, Volodarskaya S I, Novitskaya E V, Perov A N, Karasik V E, Yurchenko S O 2014 *J. Phys. Conf. Ser.* **486** 012014
[31] Taylor Z D, Singh R S, Culjat M O, Suen J Y, Grundfest W S, Lee H, Brown E R 2008 *Opt. Lett.* **33** 1258–60
[32] Arbab M H, Dickey T C, Winebrenner D P, Chen A, Klein M B, Mourad P D 2011 *Biomed. Opt. Exp.* **2** 2339–47
[33] Ashworth P C, Pickwell-MacPherson E, Provenzano E, Pinder S E, Purushotham A D, Pepper M, Wallace V P 2009 *Opt. Exp.* **17** 12444–54
[34] Pickwell E, Wallace V P, Cole B E, Ali S, Longbottom C, Lynch R J M, Pepper M 2007 *Caries Research* **41** 49-55

[35] Reid C B, Fitzgerald A, Reese G, Goldin R, Tekkis P, O'Kelly P S, Pickwell-MacPherson E, Gibson A P, Wallace V P 2011 *Physics in Medicine and Biology* **56** 4333-53
[36] Bennett D B, Taylor Z D, Tewari P, Singh R S, Culjat M O, Grundfest W S, Sassoon D J, Johnson R D, Hubschman J-P, Brown E R 2011 *J. Biomed. Opt.* **16** 057003
[37] Jacobs E, Appleby R, Prather D 2012 *Opt. Eng.* **51** 091601
[38] Furxhi O, Jacobs E L, Preza C 2012 *Opt. Eng.* **51** 091612
[39] Perov A N, Zaytsev K I, Fokina I N, Karasik V E, Yakovlev E V, Yurchenko S O 2014 *J. Phys. Conf. Ser.* **486** 012027
[40] Karpowicz N, Dawes D, Perry M J, Zhang X-C 2006 *Proc. SPIE* **6212** 62120G
[41] Stoik C D, Bohn M J, Blackshire J L 2008 *Opt. Exp.* **16** 17039–51
[42] Yakovlev E V, Zaytsev K I, Fokina I N, Karasik V E, Yurchenko S O 2014 *J. Phys. Conf. Ser.* **486** 012008

Journal of Physics: Conference Series **584** (2015) 012006 doi:10.1088/1742-6596/584/1/012006

Investigation of the spatial structure of a high-power fiber laser beam

Pavel A Nosov, Alexander F Shirankov, Alexander G Grigoryants, Roman S Tret'yakov
Bauman Moscow State Technical University, 2nd Baumanskaya str., 5-1, 105005, Moscow, Russia

E-mail: pan@bmstu.ru

Abstract. Analysis of experimental results and theoretical studies has confirmed that the free-space propagation and the transformation of a high-power fibre laser beam by an optical system can be described by the theory of laser optics. The need to use the theory of laser optics to develop modern high-performance optical systems of laser processing heads, which form the beam of the high-power fibre laser, is discussed.

1. Introduction

Laser technology widely uses fibre lasers with output radiation ranging from several kilowatts to tens of kilowatts in continuous-wave (CW) mode. These lasers are not single-mode and have a divergence of output radiation that considerably exceeds the diffraction limit [1-5].

Spatial parameters of the laser beam in the area of material processing are largely determined by the efficiency of processing operation and the implementation of its various modes. Therefore, use of an optical system that forms a laser beam with the required spatial parameters in the material processing area is imperative. Development of laser-optical systems for forming beams of coherent radiation is based on the theory of laser optics [6-9].

Powerful fibre lasers have a modular design. These lasers consist of a set of parallel active single-mode waveguides arranged within a multimode waveguide. The number of single-mode waveguides is proportional to the radiation power. For example, at a power of 4 kW there are seven single-mode waveguides. Each active single-mode waveguide radiates independently. For spatial mixing, the radiation from all single-mode waveguides is injected into a multimode waveguide called transport waveguide [1, 4, 5].

The principle of forming the radiation of high-power fibre lasers does not enable us to answer whether the propagation of a high-power fibre laser beam in free space and through an optical system can be described by the theory of laser or classical optics. Without an answer to this question, it is impossible to begin the development of modern high-performance optical systems for laser processing heads with high-power fibre lasers.

In this study, the ability to describe a high-power fibre laser beam with the parameters of the Gaussian beam, as well as the applicability of the theory of laser optics for describing the free space propagation of the beam and its transformation by an optical system, are investigated.

Content from this work may be used under the terms of the Creative Commons Attribution 3.0 licence. Any further distribution of this work must maintain attribution to the author(s) and the title of the work, journal citation and DOI.
Published under licence by IOP Publishing Ltd

2. Experimental determination of the spatial parameters of a high-power ytterbium fibre laser beam

Most studies on fibre lasers consider their configurations to be characterized only by different scales of output power, as well as by the time and spectral performances of the output radiation [1-3]. The difficulty in describing the propagation of a beam of high-power fibre laser in free space and through the optical system is considered in a few studies that mainly deal with theoretical consideration and the results of numerical simulations [10, 11]. For example, Peñano J and others [11] analyse the effects arising during the propagation of a high power laser beam through the elements of an optical system and give the formulae for calculating the beam quality (parameter M^2), which takes into account the geometrical and thermal distortion of the lens optical system.

Experiments were performed with an ytterbium fibre laser of the LS-4-K series manufactured by LLC NTO "IRE-Polus" (with a maximum output power of 4 kW in CW mode and wavelength of 1070 nm). For a laser with fibre diameters of 50 μm and 100 μm, the spatial distribution of the radiation power density along the axis of the beam passing through the following components of the optical system of the laser processing head (manufactured by the German company Precitec) was measured:

- the collimating component (COL CO 30C F125 with $f' = 125$ mm, two-lens, anti-reflection coating for 1064 nm);
- the focusing component for the 50 μm fibre (YK52 with $f' = 120$ mm, two-lens, anti-reflection coating for 1064 nm);
- the focusing component for the 100 μm fibre (YC50 with $f' = 200$ mm, single-lens, anti-reflection coating for 1064 nm).

The distribution of the radiation power density was recorded by a Prometec LASERSCOPE UFF 100 laser diagnostics system at different sections of the beam. The obtained distributions of power density were used for determining the size (semi-diameter) of the laser beam by two criteria: 1) the method of moments and 2) the size of the zone concentrating 86.5% of the total flux of the laser beam. Because these criteria gave practically identical results after measurement processing, only the second criterion was used to determine the beam size in this study.

Based on the results of beam size measurements in different sections of the laser, the following spatial parameters of the laser beam were determined: the waist size, the waist position, the confocal parameter, the angular divergence, and the beam parameter product (*BPP*).

Measurement processing can be accomplished through various methods; for example, by the least squares method. When using this method, the results of the beam size measurement should be approximated by a hyperbolic function of the following type [12]:

$$h(z) = \sqrt{a + bz + cz^2}, \tag{1}$$

where a, b, and c are coefficients, given by:

$$\begin{pmatrix} a \\ b \\ c \end{pmatrix} = \begin{pmatrix} N & \sum_{i=1}^{N} z_i & \sum_{i=1}^{N} z_i^2 \\ \sum_{i=1}^{N} z_i & \sum_{i=1}^{N} z_i^2 & \sum_{i=1}^{N} z_i^3 \\ \sum_{i=1}^{N} z_i^2 & \sum_{i=1}^{N} z_i^3 & \sum_{i=1}^{N} z_i^4 \end{pmatrix}^{-1} \begin{pmatrix} \sum_{i=1}^{N} h_i^2 \\ \sum_{i=1}^{N} z_i h_i^2 \\ \sum_{i=1}^{N} z_i^2 h_i^2 \end{pmatrix}. \tag{2}$$

Here h_i is the size (semi-diameter) of the beam on the section with the coordinate z_i, and N is the number of sections in which the power density distribution was registered $(N \geq 3)$.

The coefficients *a*, *b*, and *c* were determined and used for calculating the following beam spatial parameters:

- waist size: $h_p = \dfrac{\sqrt{4ac - b^2}}{2\sqrt{c}}$;
- waist position relative to the selected reference plane (taking the law of signs adopted in optics into account): $s_p = -\dfrac{b}{2c}$;
- confocal parameter: $z_k = \dfrac{\sqrt{4ac - b^2}}{2c}$;
- angular divergence: $2\theta = 2\sqrt{c}$;
- BPP: $BPP = \dfrac{\sqrt{4ac - b^2}}{2}$.

The angular divergence and the confocal parameters of the input beam (at the fibre output) were calculated by the formulae [1, 13]:

$$2\theta = 2\frac{M^2\lambda}{\pi h_p} = 2\frac{BPP}{h_p}, \qquad z_k = \frac{\pi h_p^2}{M^2\lambda} = \frac{h_p^2}{BPP}, \tag{3}$$

where h_p is a semi-diameter of the fibre core. For a fibre laser with a fibre diameter of 50 μm and $BPP = 2$ mm·mrad (certified value), the angular divergence and the confocal parameter are: $2\theta = 0.16$ rad and $z_k = 0.31$ mm.

According to measurement results at a fibre diameter of 50 μm, the beam size on the collimating component is equal to $h_{\text{col}} = 8.6$ mm (at 86.5% of the total flux). As the confocal parameter of the beam at the output of the fibre is small ($z_k \ll 1$ mm), the beam size is given by [6]: $h(z) = \dfrac{h_p}{z_k} \cdot |z - s_p| = \theta \cdot |z - s_p|$. The distance s from the fibre end to the first surface of the collimating component is given by: $s = \dfrac{h_{\text{col}}}{\theta}$; its value is equal to: $s = 107.5$ mm for a fibre diameter of 50 μm. As the front focus of the collimating component COL CO 30C F125 is located at a distance $s_{F\ \text{col}} = -111.33$ mm from its first surface, the distance from the front focus to the fibre end (waist position at the optical system input) is equal to: $z_{p0} = -s_{F\ \text{col}} - s = 3.83$ mm.

The spatial structure of the beam after the collimating component was also investigated. The measurements were taken for several values of radiation power (600, 900, 1200, 1800, and 2700 W) from the initial position to a distance of 100 mm, in 10 mm increments. The minimum possible distance from the last surface of the collimating component to the measurement device was equal to 45 mm. At a laser power of 2700 W and fibre diameter of 50 μm, the beam size was 9.28 mm at the initial measurement position and 9.36 mm at the final position. The power density distribution in the back focal plane of the collimating component was also recorded. The beam size in this plane was $h_{F'} = 9.34$ mm. On the basis of the principle of similitude of the optical fields in optically conjugate planes [6], the measured beam size at the back focal plane of the collimating component allows for the calculation of the angular divergence of the beam at the fibre output. Moreover, when using the 50 μm fibre, the angular divergence of the beam was $2\theta = 0.15$ rad. It was observed that the values of the angular divergence of the fibre laser beam, calculated with both the certified data and the measurement data, practically coincide.

Figure 1 (a) shows the measured values for the beam size at the collimating component output (dotted curve) and an approximating hyperbolic function (solid curve), at a laser power of 2700 W and fibre diameter of 50 μm.

To check the validity of laser optics laws in describing the transformation of the beam of the high-power fibre laser by an optical system, power density distribution at the output of the collimating component, defocused in the longitudinal direction by $z_{\text{def}} = 15.2$ mm, were measured. In this case, the fibre end was located at a distance $z_p = z_{p0} - z_{\text{def}} = -11.37$ mm from the front focus of the collimating component. Displacement of the collimating component allowed to form a converging beam [Figure 1 (b)]. These were measured from the initial position to a distance of 70 mm for several values of radiation power (600, 900, 1200, 1800, and 2700 W).

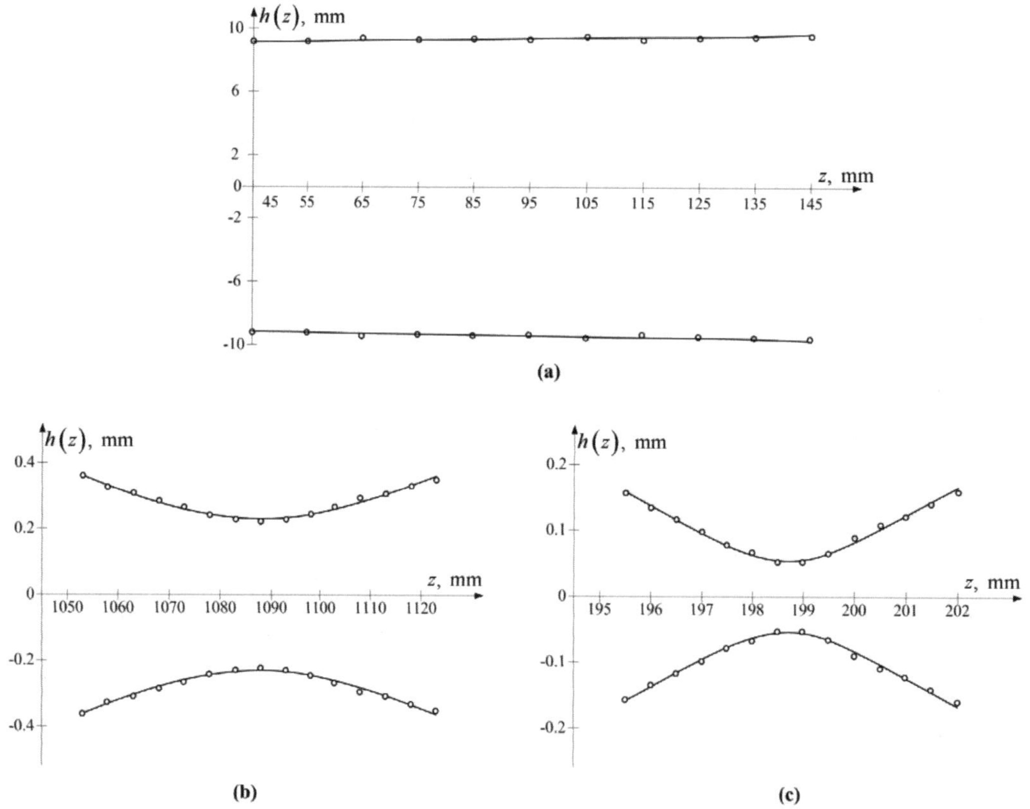

Figure 1. Beam size measurements of the laser processing head and their approximation (at a fibre diameter of 50 μm): a) at the collimating component output, b) at the defocused collimating component output, and c) at the focusing component output.

The spatial parameters of the transformed beam were calculated from the processed measurements for the case of defocused collimating component. With a laser power of 2700 W and fibre diameter of 50 μm, the spatial parameters have the following values [Figure 1 (b)]: the waist size, $h'_p = 0.23$ mm; the distance from the last surface of defocused collimating component to waist position, $s'_p = 1088.0$

mm; the distance from the back focus of the defocused collimating component to waist position, $z'_p = 963.30$ mm; the confocal parameter, $z'_k = 28.69$ mm; the angular divergence, $2\theta' = 16$ mrad; the $BPP = 1.83$ mm·mrad; and the parameter $M^2 = 5.38$.

The longitudinal magnification of a laser optical system is given by [6, 13]:

$$\alpha = \left(\frac{h'_p}{h_p}\right)^2 = \left(\frac{2\theta}{2\theta'}\right)^2 = \frac{z'_k}{z_k} = -\frac{z'_p}{z_p}. \quad (4)$$

The longitudinal magnification of the defocused collimating component had the following values, which were calculated by different beam parameters: $\alpha_{h_p} = 83.9$, $\alpha_\theta = 87.2$, $\alpha_{z_k} = 91.8$, $\alpha_{z_p} = 84.7$.

The power distribution measurements were taken for a radiation power of 600 W after focusing the components. Figure 1 (c) shows the measured values of the beam size (dotted curve) and the approximating hyperbolic function (solid curve). As a result of measurement processing, all the spatial parameters of the laser beam at the focusing component output were calculated. At a fibre diameter of 50 μm, the spatial parameters had the following values: the waist size, $h'_p = 0.054$ mm; the distance from the last surface of focusing component to waist position, $s'_p = 198.68$ mm; the confocal parameter, $z'_k = 1.13$ mm; the angular divergence, $2\theta' = 0.095$ rad; the $BPP = 2.57$ mm·mrad; and the parameter $M^2 = 7.5$.

The effect of the protective glass of the laser processing head on the parameters of the formed beam was also studied. The results obtained allowed us to conclude that the protective glass had a negligible effect on the spatial parameters and the quality of the laser beam formed in the processing area.

3. Conclusion

The multi-mode radiation of a high-power fibre laser can be described by the theory of laser optics, i.e. its free space propagation and transformation by an optical system can be described by the following spatial parameters of the laser beam: the waist size at 86.5% of the total flux, the waist position, the confocal parameter, the angular divergence. Moreover, the *BPP* of the fibre laser beam, calculated by the measurement data, and the certified *BPP* value were found to practically coincide (with less than 10% error).

Development of the optical system of laser processing head intended for forming the beam of a high-power fibre laser, should be carried out on the basis of the theory of laser optics and by recognizing that the optical elements of such a system belong to the field of power optics owing to their high power of radiation [6-9].

Acknowledgments

The study was supported by the grant of the President of Russian Federation, Project No. MK-4223.2014.8.

References

[1] Zervas M N 2014 *Int. J. Mod. Phys. B.* **28** 1442009
[2] Kurkov A S, Dianov E M 2004 *Quantum Electronics* **34** 881
[3] Dianov E M 2010 *Quantum Electronics* **40** 1
[4] Okhotnikov O G 2012 *Fiber Lasers* http://onlinelibrary.wiley.com/book/10.1002/9783527648641
[5] Grigorjants A G, Vasiltsov V V 2012 *Engineering Journal: Science and Innovation* **6** 1
[6] Pakhomov I I and Tsibulya A B 1988 Computational Methods for Laser Optical System Design *J. Sov. Laser Research* **9** 321

[7] Anikanov A G *et al.* 2010 *J. Opt. Technol.* **77** 101
[8] Pakhomov I I *et al.* 2010 *J. Opt. Technol.* **77** 107
[9] Nosov P A *et al.* 2011 *J. Opt. Technol.* **78** 586
[10] Mohring B *et al.* 2013 *Proc. SPIE* **8733** 873305
[11] Peñano J *et al.* 2009 *J. Opt. Soc. Am. B* **26** 503
[12] International standard ISO 11146-1: Lasers and laser-related equipment – Test methods for laser beam widths, divergence angles and propagation ratios – Part 1: Stigmatic and simple astigmatic beams (first edition 2005-01-15)
[13] Johnston Jr T F 1998 *Appl. Opt.* **37** 4840

Improved technique for picosecond pulse duration measurement based on second harmonic generation

D A Shelestov*, V E Karasik, K I Koshelev, V A Lazarev and A B Pnev

Bauman Moscow State Technical University, 2nd Baumanskaya str. 5/1, 105005, Moscow, Russia

E-mail: *dima.shelestov@gmail.com

Abstract. In present paper we demonstrated an improved technique for characterization of ultra-short pulse duration (PD) based on second harmonic generation (SHG). We utilized Yb-doped fiber laser with picosecond duration of pulses as implementation of proposed technique and for estimating the accuracy of pulse duration measurement. $KTiOPO_4$ crystal was used for the purpose of second harmonic generation. We determined the duration of laser pulses by analysing the combination of signal intensities on input and output of SHG crystal. We observed experimentally error of pulse duration measurement less then 2%. We showed strong dependence between the average power of SHG and the duration of picosecond pulses.

1. Introduction and Background

In technological field ultrashort-pulsed (USP) lasers are actively used in micromachining, eye surgery, optical coherence tomography etc. [1, 2] Using of USP lasers as frequency divider to organize precise link between radio and optical frequencies is one of the important scientific applications [3, 4, 5, 6, 7, 8]. Wide spreading in industry requires better parameters stability: average power, pulse duration (PD), position of polarization plane, mode of laser generation. Commonly used average power stability systems are typical for modern CW lasers. USP lasers would use these systems also but power affecting on PD require additional subsystems. During long-term utilizing fiber USP lasers demonstrate undesirable PD shifts, which could lead to critical failure of the target process. Undesirable PD shifts could be detected and compensated by an embedded PD measurement system.

Among the various PD measurement methods, we can highlight the following as most popular: streak camera, intensity autocorrelator [9], interferometric autocorrelator (or fringe-resolved autocorrelation - FRAC) [10, 11], frequency-resolved optical gating (FROG) [12], GRENOUILLE [13], and spectral phase interferometry for direct electric-field reconstruction (SPIDER) [14]. PD measurement based on SHG effect is less accurate then last versions of mentioned methods, nevertheless we did a research to understand the maximum possible accuracy. Balance between simplicity and accuracy of improved SHG method makes this technique useful for practical applications.

Previous papers dedicated to SHG generation [15] [16] had no ideas to use SHG as a measurement method. Harmonic generation process for ultrashort pulses was well studied in [17, 18]. Experiment setup used in [19] was similar to our setup, but we utilize USP laser. Slow

Content from this work may be used under the terms of the Creative Commons Attribution 3.0 licence. Any further distribution of this work must maintain attribution to the author(s) and the title of the work, journal citation and DOI.
Published under licence by IOP Publishing Ltd

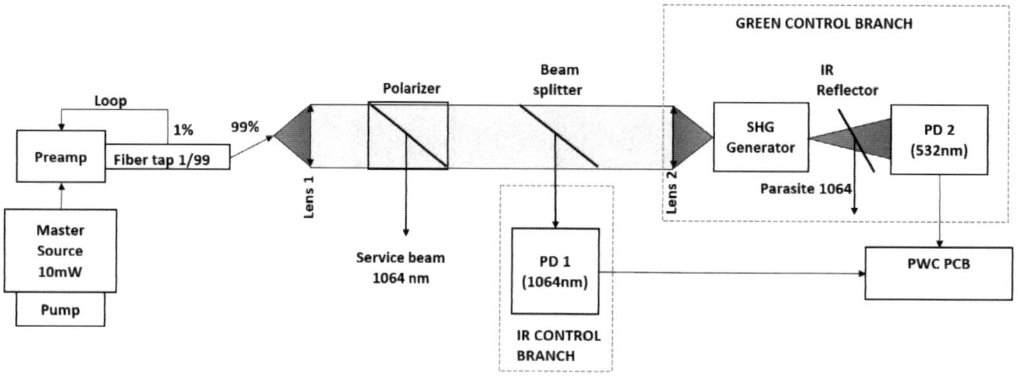

Figure 1. PD measurement setup

photodiodes did not detect ultra-shot processes, however our technique of analyzing CW-signal leaded to improvement in accuracy.

2. Experiments

During long-term use of ultra-short laser we observed undesired PD shifts, which cause time-consuming equipment returns and corresponding financial losses for laser development companies. Detected shifts have unclear reasons, but they all are characterized by a time constant greater than 1 s.

The general idea of PD measurement by means of SHG is to simultaneously fix average intensity immediately before $I_{av}^{(1)}$ and after $I_{av}^{(2)}$ SHG crystal and to calculate the current PD value by use of measured intensities and initial calibration data.

Setup consisted of Yb-doped fiber laser (Master source) followed by fiber amplifier (Preamp) (figure.1). SMF-128 fiber fragment compensated dispersion induced by amplifier. Then radiation came to polarizer and beam-splitter, where light partly redirected to photodiode 1, but main intensity run to SHG crystal. Photodiode 2 fixed output SHG signal after filtering. Overall conversion efficiency in our setup was less than 0.1% , although efficiency of KTiOPO4 nonlinear crystal itself was about 1%. For more productive nonlinear elements, such as optimized crystals [20] or globular photonic crystals [21, 22], our method precision will rise.

The setup ensured that the PD after amplification was consistent with the laser output. Rotation of the polarization plane had a strong influence on SHG efficiency, as described by $I_{av}^{(2)} \sim \sin(2\varphi)$ where $I_{av}^{(2)}$ is the average intensity of the SHG signal and φ is an angle between the incident light polarization plane and the crystal optical axis. To avoid this effect, a polarizer was used to convert the oscillation of the angular position of the polarization plane into average power oscillation. Since InGaAs photodiodes detected both 532 nm and 1064 nm, even 1% of the incident 1064 nm radiation entering the second photodiode was several times stronger than the SHG signal. An IR reflector passed the 532 nm beam through and deflected the parasitic beam. A special electronic board collects signals from both photodiodes simultaneously.

3. Results

By means of linear approximation, we determined the relationship between PD and SHG signal for constant 1064 nm radiation power. Amplifier closed loop controlled average intensity on the

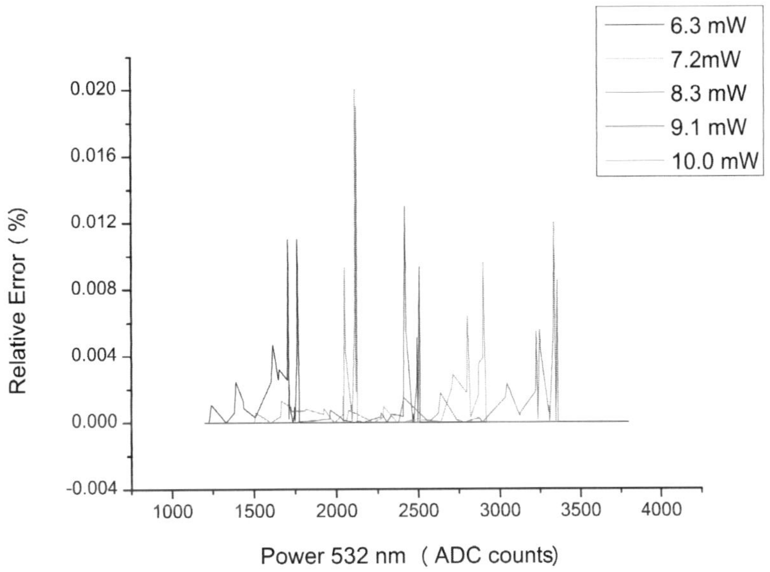

Figure 2. Relative error between estimated and real PD (Each line means constant average power 1064 nm radiation)

SHG crystal input 1 by means of photodiode with time constant T_{pd}.

$$I_{av}^{(1)} = \int_0^{T_{pd}} I^{(1)}(t)dt = const \tag{1}$$

Dependance of SHG average intensity on PD was described by

$$I_{av}^{(2)}(\tau) \sim \frac{1}{\tau^2} \tag{2}$$

We fixed new series with $G(\tau)$ representing a linear interpolation of SHG dependance on PD. We got each curve from one $I_{av}^{(1)}$ value from the range $R = (6.3mW; 8.3mW; 10.0mW)$. We found that there is a similarity in the form of the calculated curves for different $I_{av}^{(1)}$ values. After normalization, this relationship can be expressed as

$$k(\tau) = \frac{1}{j}\sum_{j=1}^{N} \frac{G_j}{max[G_j]} \tag{3}$$

The mismatch between the series of curves $G_j(\tau)$ and $k(\tau)$ was less than 0.01%. For better results $k_2(y)$ calibrations were measured on the upper limit of the SHG signal range. We found that the relative error while using $k(\tau)$ is 5 times less than in the interpolation calculation, with a maximum of 2% (figure.2).

4. Conclusion

We improved the simple PD measurement method based on SHG effect. We revealed clear relation between SHG average power level and PD. By means of utilizing of observed relation in PD processing we reached 2% accuracy level.

Acknowledgement
The authors would like to thank S. Yurchenko for valuable comments.

References
[1] Fermann M and Hartl I 2013 *Nature Photonics* **7** 868–874
[2] Fermann M, Galvanauskas A and Sucha G 2003 *Ultrafast Lasers: Technology and Applications* 1st ed (Basel: Marcel Dekker AG)
[3] Tausenev A V, Kryukov P G, Bubnov M M *et al.* 2005 *Quantum Electron.* **35** 581
[4] Gubin M A, Kireev A N, Kozlovskii V I *et al.* 2012 *Quantum Electron.* **42** 509
[5] Gubin M A, Kireev A N, Kozlovskii V I, Lazarev V A *et al.* 2012 *Quantum Electron.* **42** 565
[6] Tarabrin M, Frolov M, Gubin M, Kireev A *et al.* 2014 Testing of cr:znse laser with intracavity methane cell in 77–300 k temperature range *Proceedings of Laser Optics 2014* pp 1–1
[7] Lazarev V A, Sazonkin S S, Pniov A B *et al.* 2014 *JPCS* **486** 012004
[8] Tsapenko K P, Lazarev V A, Leonov S O and Pnev A B 2014 *JPCS* **486** 012006
[9] Armstrong J A 1967 *Appl. Phys. Lett.* **10** 16–18
[10] Hache F, Driscoll T J, Cavallari M and Gale G M 1996 *Appl. Opt.* **35** 3230–3236
[11] Ouyang X, Ma J, Yang L *et al.* 2012 *Appl. Opt.* **51** 3989–3994
[12] Kane D and Trebino R 1993 *IEEE J. Quantum Electron.* **29** 571–579
[13] O'Shea P, Kimmel M, Gu X and Trebino R 2001 *Opt. Lett.* **26** 932–934
[14] Gallmann L, Sutter D H, Matuschek N, Keller U *et al.* 1999 *Opt. Lett.* **24** 1314–1316
[15] Saltiel S, Savov S and Tomov I 1982 *Opt. and Quantum Electron.* **14** 391–394
[16] Franken P A, Hill A E, Peters C W and Weinreich G 1961 *Phys. Rev. Lett.* **7**(4) 118–119
[17] Glenn W 1969 *IEEE J. Quantum Electron.* **5** 284–290
[18] Chou H F, Lin C F and Wang G C 1998 *J. Lightw. Technol.* **16** 1686–1693
[19] Smith W and Bechtel J 1976 *J. Appl. Phys.* **47** 1065–1068
[20] Cussat-Blanc S, Ivanov A, Lupinski D and Freysz E 2000 *Appl. Phys. B* **70** S247–S252
[21] Zaytsev K I, Katyba G M, Yakovlev E V, Gorelik V S and Yurchenko S O 2014 *J. Appl. Phys.* **115** 213505
[22] Zaytsev K I and Yurchenko S O 2014 *Appl. Phys. Lett.* **105** 051902

On accumulated signal degradation in a cascade of semiconductor optical amplifiers

N D Moroz*[1], G de Valicourt[2], A Ghazisaeidi[2], J-C Antona[2]
[1] Laser and Optoelectronic Systems Department, Bauman Moscow State Technical University, 105005, 2nd Baumanskaya street 5, Moscow, Russia.
[2] WDM Dynamic Networks Department, Alcatel-Lucent Bell Labs, Nozay 91620, France

E-mail: *ndmoroz@gmail.com

Abstract. We study a design and limitations of a wavelength division multiplexing packet-switched network based on semiconductor optical amplifiers (SOAs) under different modulations formats (QPSK, 8-QAM and 16-QAM). A simple convergence rule for accumulated SOA nonlinearities based on the so-called nonlinear threshold is proposed. We calculate the signal reach depending on various SOA and network parameters by means of numerical simulation. We provide a design tool enabling optimization of a SOA-based optical network architecture.

1. Introduction

Emerging bandwidth-on-demand services in optical fiber networks induce new distributed and bursty traffic profiles. Therefore optical packet-switched metropolitan networks have been proposed [1]. In this scenario, optical packets occupy time slots and slot blocker (SB) device either routes or blocks packets of each wavelength in the network. We anticipate semiconductor optical amplifier (SOA) as a promising candidate for the SB, given that it can operate as a fast modulator with sufficient optical gain and high extinction ratio [2]. This scenario implies one SB per wavelength, thus, the SOA is used only in a single-channel configuration.

The SOA limits the signal reach due to the optical signal-to-noise ratio (OSNR) degradation by amplified spontaneous emission (ASE) and nonlinear distortions in the passing signal, mainly self-phase modulation (SPM). An analytical model enabling to characterize the accumulation of SOA nonlinearities over a 10G on-off keying non-return-to-zero (OOK NRZ) signal has been proposed and validated [3].

In this paper, we extend this study to advanced modulation formats such as M-ary quadrature amplitude modulation (M-QAM), to SOAs with various parameters, and to optical fiber-SOA cascades. The capability of SOA to amplify complex modulation formats has been numerically and experimentally investigated for one device [4]. Numerical simulations have been proposed to study the impact of the in-line SOA nonlinear penalty including the channel count and the cumulated chromatic dispersion (CD) [5]. However, general design rules including the accumulation of nonlinear distortions along the SOA cascade are still missing. Here, we establish a simple model to derive and verify rules for cascaded SOAs nonlinearity for advanced modulation formats, various inter-SOA fiber lengths and

Content from this work may be used under the terms of the Creative Commons Attribution 3.0 licence. Any further distribution of this work must maintain attribution to the author(s) and the title of the work, journal citation and DOI.
Published under licence by IOP Publishing Ltd

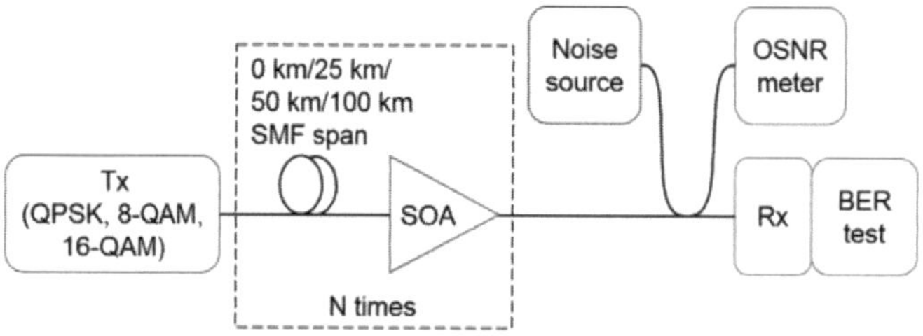

Figure 1. The simulation setup for SOA penalties investigation; Tx — transmitter, Rx — receiver.

various SOA parameters. Using these accumulation rules and the cumulated OSNR degradation, we assess the maximum number of cascaded SOAs while guaranteeing a certain quality of transmission.

2. Simulation setup

We simulate an optical fiber line presented in Figure 1. It depicts a 28 Gbaud optical signal path through a cascade of SOAs in a prospective fiber network. We use root-raised cosine filter with 0.1 roll-off for pulse shaping a payload of 2^{15} symbols.

Our SOA model has 14.2 dB unsaturated gain, saturation power of 12 dBm, noise figure of 8 dB, linewidth factor of 5, and 100 ps charge carrier lifetime. We also implement propagation loss of 0.05 respectively to the gain. The implemented space-resolved model with 10 sections for SOA simulation is described in [6].

We use our setup for three types of simulations. First, SOA parameters are fixed and we simulate a cascade of SOAs with three signal modulation formats: quadrature phase-shift keying (QPSK), 8-QAM, and 16-QAM. Second, we simulate a cascade of 25 km, 50 km and 100 km spans of single-mode fiber (SMF) by one SOA and 100G QPSK signal is sent over such network simulation setup. Third, we use 100G QPSK signal and a cascade of SOAs varying their linewidth factor from 3 to 6 and carrier lifetime from 100 ps to 400 ps.

We vary optical power at the input of the first SOA and maintain the same input power for each SOA in the cascade. For the simulation setup with a cascade of optical fiber spans and SOAs input power of -15 dBm corresponds to the launched power into the SMF of -10 dBm (25 km), 5 dBm (50 km), and 5 dBm (100 km).

To evaluate OSNR degradation we measure its level right before the reception. To assess nonlinear distortions we calculate OSNR penalties loading ASE-type complex white Gaussian noise at the receiver input. Noise loading keeps a constant BER = 10^{-3} and we calculate the OSNR penalty as the difference between the required OSNRs (at BER = 10^{-3}) after the SOA cascade and in a back-to-back (B2B) configuration.

Varying both input optical power P_{in} and number of devices N in the cascade we consider OSNR penalty as a function of the product $P_{in}\times N$, referred to as an integrated power (IP) in [3]. By analogy with [3], we also define the nonlinear threshold (NLT) as the minimum value of IP leading to a given threshold penalty. In the case of nonlinear noise occurring in fiber networks without inline CD compensation, the threshold OSNR penalty equals 1.76 dB providing the best transmission quality [7]. We use the same penalty to assess the NLT after the cascade of SOAs.

3. Simulation results and accumulation rules

We investigate the convergence of OSNR penalty curves depending on the IP for different modulation formats in figure 2(a) and in absence of fiber spans. The modeled OSNR penalty is plotted against the IP (in dB scale, i.e. $10\log10[P_{in,mW}\times N/1\ mW]$). Convergence to a single curve makes it tempting to approximate the penalty as a function of the IP $P_{in}\times N$ as previously described for OOK signals [3].

For QPSK, we obtain a maximum IP spread of 2.6 dB (1.5 dB after passing through 3 SOAs). Moving to M-QAM formats, the OSNR penalties is also plotted depending on the IP and we achieve less than 1.5 dB overall spreads for OSNR penalties with 8-QAM and 16-QAM signals. Obtained

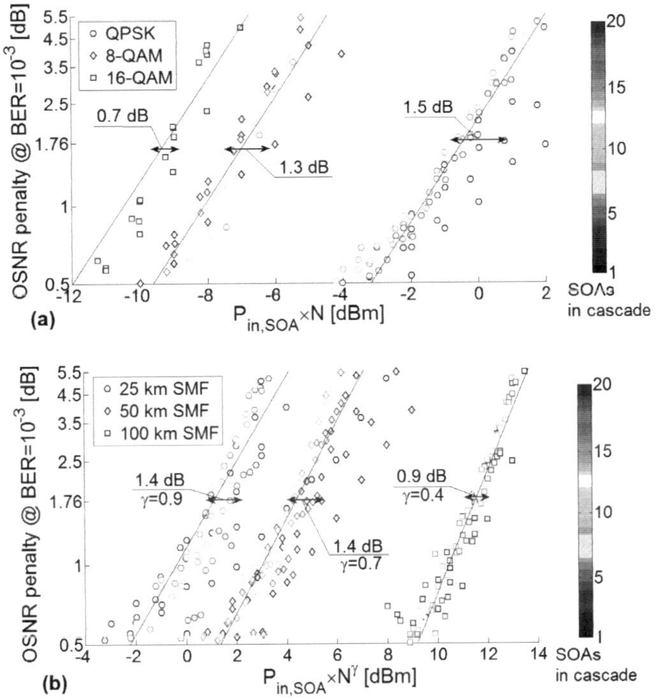

Figure 2. OSNR penalties for a cascade of 20 SOAs with QPSK, 8-QAM and 16-QAM signal (a) and for a cascade of 20 SOAs with 25 km, 50 km and 100 km SMF spans (b).

spread of NLT powers is acceptable for all modulation formats under study (spread of 1.6 dB was obtained with OOK signals [3]).

With a cascade of SOAs and fibers, we found that the basic IP scale produce the spread of 2 dB for 50 km spans and 4 dB for 100 km spans. Moreover, the accumulation of nonlinearities tends to change its character when the fiber length changes, so the scale has to be dynamic. Motivation to get the spread of 1.5 dB or less and study of this accumulation character brought us to a more general law – modified IP $P_{in}\times N^{\gamma}$ that we implement in Figure 2(b). Varying the exponent γ, we take into account effects arising in the fiber-SOA cascade: different amounts of cumulated fiber nonlinearities, an interaction between nonlinearities of the fiber and the SOA and an interaction of CD with the SOA nonlinearity.

4. Discussion

The largest count of cascadable devices is obtained at the optimal trade-off between cumulated nonlinear distortions and ASE noise. Both phenomena accumulate from one device to another. We estimate the ASE noise limit with the SOA input power that provides a certain OSNR: the OSNR in B2B for BER=10^{-3} plus 4.5 dB penalty (2.7 dB assumed for aging and implementation penalties as well as 1.8 dB coming from the non-linear threshold). B2B OSNR at BER=10^{-3} are the following: 13.1 dB for QPSK signal, 17.2 dB for 8 QAM signal, and 20.1 dB for 16-QAM signal. The NLT and noise limiting power values against the number of SOA are displayed for different signal modulations in figure 3 (a) and different fiber span lengths in figure 3(b); the impact of the linewidth factor and the carriers' lifetime of the SOA on signal degradation – in figure 3(c) and figure 3(d).

We propose to use these results as a design tool for SOA-based optical network in the following way. In figure 3 NLT curves characterize maximum tolerable nonlinearity and OSNR curves limit acceptable ASE noise level. Therefore, the maximum number of cascaded devices can be found at the

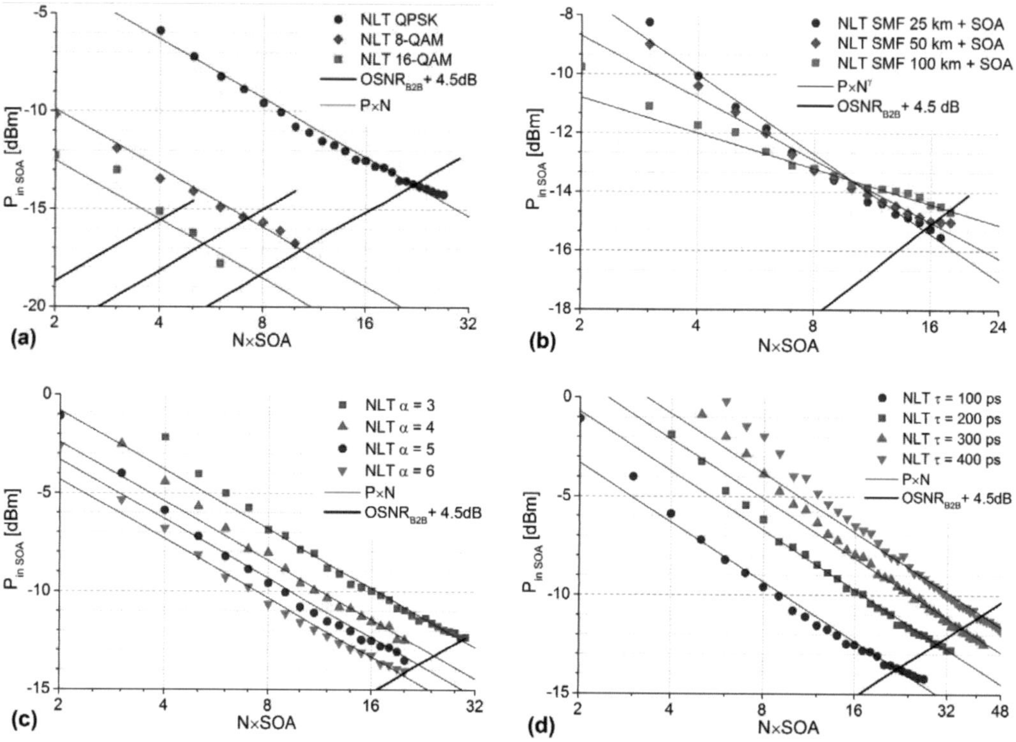

Figure 3. NLT and OSNR limiting power versus number of cascaded SOAs for QPSK, 8-QAM, 16-QAM formats of signal (a), for QPSK signal and 25 km (γ=0.9), 50 km (γ=0.7), and 100 km (γ=0.4) SMF inter-SOA spans (b), for QPSK signal and SOA linewidth factor 3, 4, 5, 6 (c) and for QPSK signal and SOA carriers' lifetime 100 ps, 200 ps, 300 ps, and 400 ps (d).

intersection of NLT and OSNR curves. The intersection values are listed in table 1 for various network configuration and in table 2 for various SOA parameters.

The slope of the NLT curves is the same for three formats in figure 3(a) thus the nonlinear distortions have the same fundamental nature (extra phase noise added by the saturated SOAs). We explain a large shift from QPSK to M-QAMs due to the other amplitude levels appearing in 8-QAM and 16-QAM. It demonstrates extra sensitivity due to a decrease in the extinction ratio of the traversing signal (as explained for OOK signals [3]).

In figure 3(b), when different amounts of fiber and CD mix with SOA-induced distortions, the convergence exponent γ tends to change. Starting with steep γ=0.9 for 25 km span extension of the fiber part reduces the slope to γ=0.4 for 100 km. We explain this phenomenon considering different input powers in optical fiber and large amounts of cumulated CD. Induced by the fiber nonlinearity γ is close to 0.67 for 100km-long CD-unmanaged 100G systems as a result of noise dependence on span input cumulated CD and uncorrelated noise contributions from span to span due to the large relative span input cumulated CD [8]. This effect manifests for more than 11 cascaded fiber-SOAs and provides fewer penalties for longer spans.

Increasing the linewidth factor and decreasing the carriers' lifetime make the SOA more nonlinear due to intensified SPM and patterning effect respectively. Found results in Figure 3(c) and Figure 3(d) conform to this reasoning. We observe the same slope for all the cases and apply a common $P_{in}\times N$ convergence rule to demonstrate identical nature of the penalties. As linewidth factor variations are less influential for high degree M-QAM formats as concluded in [4], SOA with shorter carrier lifetime should be chosen for complex modulation signals.

Table 1. Maximum number of SOAs in various network setups

Network configuration	N×SOA
QPSK	22
8-QAM	7
16-QAM	4
QPSK + 25 km SMF	15
QPSK + 50 km SMF	16
QPSK + 100 km SMF	17

Table 2. Maximum number of SOAs for various SOA parameters

SOA parameter and its value	N×SOA
$\alpha = 3$	29
$\alpha = 4$	24
$\alpha = 5$	22
$\alpha = 6$	19
$\tau = 100$ ps	22
$\tau = 200$ ps	29
$\tau = 300$ ps	35
$\tau = 400$ ps	41

5. Conclusion

By the means of numerical simulation, we assessed the impact of the cascade of SOAs on a network transmission varying modulation, fiber length and SOA parameters.

We verified the applicability of the integrated power rule $P_{in}\times N$ to the accumulation of nonlinearities in SOA cascade for QPSK and M-QAM signals. We proposed modified integrated power rule $P_{in}\times N^{\gamma}$ for a cascade of fibers and SOAs. Both rules provide a spread of OSNR penalties less than 1.5 dB in power scale from 3 cascaded SOAs. Generalizing nonlinearity accumulation for smaller number of SOAs or more precise prediction requires thorough analytical investigation.

Acknowledgements

N D Moroz thanks Francesco Vacondio and Vladimir Lazarev for fruitful discussions.

This work was supported in part by the FP7 SASER project.

References

[1] Chiaroni D 2010 *36th European Conference and Exhibition on Optical Communication (ECOC 2010)* 1–12

[2] De Valicourt G, Moroz N D, Jenneve P, Vacondio F, Duan G H, Jany C, Leliepvre A, Accard A and Antona J C 2014 *IEEE Photonics Technology Letters* **26** 678–81

[3] De Valicourt G, Simonneau C, Ghazisaeidi A, Brenot R, Antona J-C and Bigo S 2013 *IEEE Photonics Technol. Lett.* **25** 2389–92

[4] Bonk R, Huber G, Vallaitis T, Koenig S, Schmogrow R, Hillerkuss D, Brenot R, Lelarge F, Duan G-H and Sygletos S 2012 *Optics Express* **20** 9657–72

[5] Bendimerad D F, Ghazisaeidi A, Vuong J, Ramantanis P, Seck A, Renaudier J and Frignac Y 2013 *39th European Conference and Exhibition on Optical Communication (ECOC 2013)* pp 1–3

[6] Ghazisaeidi A and Rusch L A 2011 *J. Lightwave Technology* **29** 3331–39

[7] Grellier E and Bononi A 2011 *Opt. Exp.* **19** 12781–88

[8] Vacondio F, Rival O, Simonneau C, Grellier E, Bononi A, Lorcy L, Antona J-C and Bigo S 2012 *Optics Express* **20** 1022–32

Nonlinear conversion in optical waveguide filled with $NaNO_2$

Irina N Dolganova, Kirill I Zaytsev, Stanislav O Leonov, Aleksandr O Schadko, Kirill A Komarov and Stanislav O Yurchenko

Bauman Moscow State Technical University, 2nd Baumanskaya str. 5, Moscow, Russia, 105005

E-mail: in.dolganova@gmail.com

Abstract. This paper contains the first experimental results of nonlinear conversion in optical waveguide filled with solid nonlinear media. The experimental sample was made from SiO_2 fiber of the 1.6 cm length. During the pumping of the sample by the pulse laser, the generation of second harmonic radiation was observed at the whole length of the sample filled with $NaNO_2$. Despite the visibility of this effect, the detected signal was rather weak due to the scattering inside the fiber. Nevertheless, the successful experiment presented in this work may lead to the future development of new nonlinear devices using photonic crystal fiber filled with solid nonlinear medium.

1. Introduction

It is rather promising to fabricate microstructures filled with nonlinear media [1], [2], [3], [4], [5], [6], [7], [8], [9]. In such media as filled photonic crystals, microcells, photonic crystal fiber [10], [11], [12], [13], nonlinear conversion can be more efficient than in bulk material [4], [7], [8]. Thus, there is a huge interest to these structures, according to recent publications (see for example [14], [15], [16], [17], [18], [19]).

Turning the attention to the hollow core optical fiber, it is possible to find the results of filling it with nonlinear liquid [1], [2], [6] or gas [3], [5], [14]. For more information about such structures see [20]. But there a lot of difficulties, primarily technological, for fabricating structures filled with solid medium. On the other hand, structures filled with solids have no disadvantages of ones filled with liquid and gas: maintaining the particular pressure for gas, vacuum camera for beam input and output [5]. At least, such solid filled nonlinear fibers can be used for making an effective coherent two-wavelength source [1], photonic microcells [20], and nonlinear radiation converter [15].

2. Sample

We have chosen $NaNO_2$ because of its remarkable properties. At first, its melting temperature is $544K^{\circ}$, which is lower than softening temperature of SiO_2. Second, at room temperature $NaNO_2$ is a powder, so it is rather easy to work with it. Third, $NaNO_2$ is a well-known nonlinear material [21]. Also it has been already used successfully for filling the nanostructured materials [4], [9]. Anisotropy of $NaNO_2$ leads to different refractive indexes $n_x = 1.3395, n_y = 1.4036, n_z = 1.6365$ [22]. In case of SiO_2 optical tube, which has $n = 1.54$ refractive index, we have to use only n_z refractive index of sodium nitrite in order to maintain the waveguide

Content from this work may be used under the terms of the Creative Commons Attribution 3.0 licence. Any further distribution of this work must maintain attribution to the author(s) and the title of the work, journal citation and DOI.
Published under licence by IOP Publishing Ltd

condition $n_{core} > n_{tube}$. Thus, it is required to obtain technologically the nonlinear crystal with particular orientation.

For our experiment we used the hollow core optical fiber of 10 cm length with internal diameter $d = 80\ \mu m$. The hollow fiber was filled with liquid $NaNO_2$ by means of using vacuum compressor. During this process, the nonlinear media filled some part of the fiber and crystallized inside. The length of the filled fiber was 1.6 cm.

3. Experiment

Figure 1 shows the generation of second harmonic generation in the sample, during pumping it with femtosecond laser pulses. Laser wavelength was $\lambda = 1.03\ \mu m$, so that the second harmonic wavelength corresponds to $\lambda = 515\ nm$. The green light is well observed from figure 1, because the green filter was used for making picture, in order to cut off the laser radiation.

Figure 1. Second harmonic generation in the sample of SiO_2 fiber filled with $NaNO_2$.

We can observe the second harmonic generation along the whole sample. This approves the fact, that the sample became a waveguide and the required crystal grown in the core. But despite the visible result, a lot of radiation was scattered in the fiber cladding. The reason is in the inhomogeneity of the grown crystal. It should be mentioned that scattering in this sample is determined by the final structure of nonlinear crystal, and the impact of the structure factor in this case is significant (for example see [23]). Nevertheless, the appropriate technological cooling process can solve this problem in the future.

4. Results

The output power at the end of the sample was measured in order to estimate the quality of nonlinear process. From figure 2 this signal spectrum can be clearly divided into the incident radiation transmitted through the fiber and second harmonic radiation. This second harmonic's peak was measured several times during changing of the pumping laser power (figure 3). The corresponding growth of the second harmonics generation is presented. The broadening of the 515 nm peack can be explained by the instability of the laser source and by the pulse broadening during the propagation through the sample.

Figure 4 shows the changes of the transmitted radiation power, when the laser pulse power was changed from 170 mW to 3.5W. It should be mentioned that the incident laser beam was focused in the core center at the top of the sample. P_{out} didn't include the radiation scattered to other direction.

The initial part of the power dependence is well approximated by the equation $P_{out} = 0.0665 \cdot P_{in}^2$. This square-dependence equation is typical for second harmonic generation.

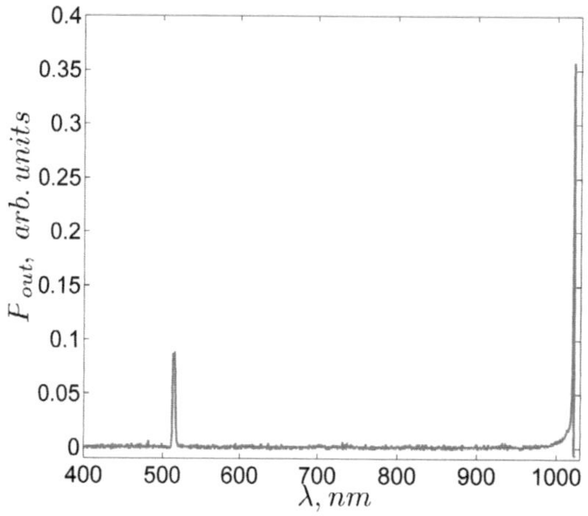

Figure 2. Detected signal spectrum with well recognized second harmonic peak.

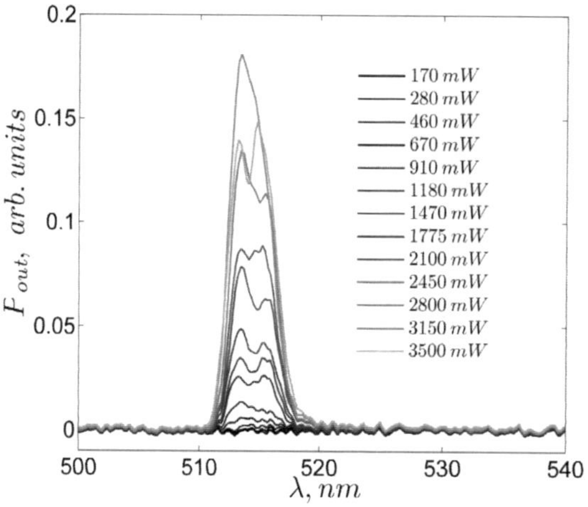

Figure 3. Transmitted signal spectrum in the region of second harmonic measured for different incident radiation power.

Nevertheless, losses in the crystalline sample core lead to the divergence between the experiment and approximation.

5. Conclusion

Second harmonic generation in the optical fiber filled with solid nonlinear media approves the possibility of the nonlinear conversion in photonic-crystal fiber where it is possible to obtain lower dispersion. Also the technological process of fiber filling with nonlinear media can be improved in order to make the nonlinear crystal in the sample core become more homogeneous. This is possible by temperature control during the crystallization process. All these means will be helpful in order to produce sample with higher effectiveness of nonlinear transformation by reducing the scattering and dispersion losses.

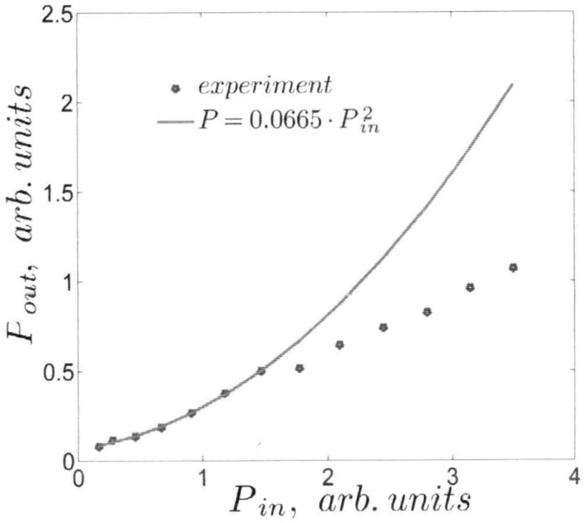

Figure 4. Transmitted power dependence on the incident laser power: experimental results and approximation equation.

Acknowledgement
This work was supported by the Russian Scientific Foundation (grant # 14-29-00277).

References

[1] Zheltikov A M and Naumov A N 2000 *Quantum Electronics* **30** 351
[2] Bartels R A, Wagner N L, Baertschy M D, Wyss J, Murnane M M, and Kapteyn H C 2003 *Opt. Lett.* **28** 346
[3] Xu Y, Chen X, and Zhu Y 2008 *Opt. Exp.* **16** 9205
[4] Belotitskii V I, Kumzerov Yu A, and Fokin A V 2008 *JETP. Lett.* **87** 465
[5] Kieu K, Schneebeli L, Norwood R A, and Peyghambarian N 2012 *Opt. Exp.* **20** 8148
[6] Midorikawa K, Tamaki Y, Itatani J, Nagata Y, Obara M 1999 *IEEE J. Selected Topics in Quant. Electr.* **5** 1475
[7] Zaytsev K I, Katyba G M, Yakovlev E V, Gorelik V S and Yurchenko S O 2014 *J. Appl. Phys* **115**
[8] Zaytsev K I, Yurchenko S O 2014 *Appl. Phys. Lett.* **105** 051902
[9] Murzina T V, Sychev F Yu, Kolmychek I A, and Aktsipetrov O A 2007 *Appl. Phys. Lett* **90** 161120
[10] Joannopoulos J D, Villeneuve P R, Fan S 1997 *Nature* **386** 143
[11] Rigby P 1998 *Nature* **396** 415
[12] Mingaleev S F, Kivshar Y S 2001 *Phys. Rev. Lett.* **86** 5474
[13] Russell P 2003 *Science* **299** 358
[14] Vieweg M, Gissibl T, Pricking S, Kuhlmey B T, Wu D C, Eggleton B J, Giessen H 2010 *Opt. Exp.* **18** 25232
[15] Khetani A, Riordon J, Tiwari V, Momenpour A, Godin M, and Anis H 2013 *Opt. Exp.* **21** 12340
[16] Benabid F, Knight J C, Antonopoulos G, and Russell P S J 2013 *Science* **298** 399
[17] Benabid F, Couny F, Knight J C, Birks T A, and Russell P S J 2005 *Nature* **434** 488
[18] Nold J, Holzer P, Joly N Y, Wong G K L, Nazarkin A, Podlipensky A, Scharrer M, and Russell P St J 2010 *Opt. Lett.* **35** 2922
[19] Travers J C, Chang W, Nold J, Joly N Y, and Russell P St J 2011 *JOSA B* **28** A11–A26
[20] Benabid F, Roberts P J, Couny F, Light P S 2009 *J. EOS* **4** 09004
[21] Vogt H and Happ H 1971 *Physica Status Solidi (b)* **44** 207
[22] Martienssen W, Warlimont H 2005 *Handbook of Condensed Matter and Materials Data* (New York: Springer)
[23] Fokina I N, Karasik V E, Orlov V M, Budak V P 2014 *J. Quant. Spectrosc. Radiat. Transfer* **149** 108

The versatile terahertz reflection and transmission spectrometer with the location of objects of researches in the horizontal plane

N S Balbekin, Ya V Grachev, S V Smirnov and V G Bespalov
ITMO University, St. Petersburg 197101, Russia
E-mail: Godnikolay@yandex.ru

Abstract. The experimental setup of versatile pulse terahertz reflection and transmission spectrometer and operation concept were described. Using this setup the temporary forms of THz signal reflected from and transmitted through the samples like amino acids, dental tissue and normal were obtained. The possibility of structural determination of powdered media was demonstrated using this method.

1. Introduction

Terahertz (THz) optics and spectroscopy, and also technologies using the most long-wave optical frequency range are gaining interest because of the common use for scientific purposes, as well as for applications in the civil and military applications. The emergence of this trend is primarily due to the appearance of sources and receivers of coherent THz radiation in the 80 years of XX century [1, 2]. During the following years, due to the development of femtosecond lasers and microelectronics, there has been considerable progress in THz region studies. New methods for the generation, distribution control and detecting of THz radiation appeared, first monographs were written [3, 4].

At present, methods of generation and detection of THz radiation using pulsed lasers in the near IR range spectrum with femtosecond duration are the most developed [5]. Such detection methods allow registering directly temporal form of the electric field of THz pulse. Complex spectrum of radiation, spectra of the transmission or reflection substances and permittivities in this frequency range is obtained using the Fourier transform [6].

Huge number of spectral singularities of substances is observed in THz frequency range, energy of THz quanta corresponds to the rotational transitions of the molecules, the vibrational modes of organic compounds, the lattice vibrations in solids, intraband transitions in semiconductors and energy gaps in superconductors. Pulsed THz time-domain spectroscopy allows to investigate all of these spectral features. Broadband THz radiation reflected from the object / transmitted through the object carries a lot of information about its internal structure and spectral properties. However, standard configurations of THz spectrometers require that the sample is placed in a vertical plane. It makes difficulties when dealing with powdered samples or liquid media. The proposed scheme of the spectrometer allows solving such problems like these. This technique of researches can be considered as an ideal (non-contact, non-ionizing,

Content from this work may be used under the terms of the Creative Commons Attribution 3.0 licence. Any further distribution of this work must maintain attribution to the author(s) and the title of the work, journal citation and DOI.
Published under licence by IOP Publishing Ltd

safe) non-destructive testing technology that provides high spatial resolution both in coatings, and in depth of the investigated sample.

2. The experimental setup

For the present paper we erected the experimental setup generating broadband pulsed THz radiation by means of photoconductive antenna (undoped InAs) [7, 8] placed in a constant magnetic field of 2.4 T [9] , which was exposed to radiation of femtosecond laser pulses FL-1 (the active medium - Yb: KYW; λ = 1040nm, t_p = 46 fs, ν = 70 MHz, P = 1,2W). The parameters of THz radiation were: spectral range from 0.05 to 1.6 THz, average power 30 mW, pulse power 755 mW, pulse duration 2.7 ps. The main power was distributed between 0.12 and 1.10 THz. The scheme of general-purpose THz spectrometer is shown on figure 1.

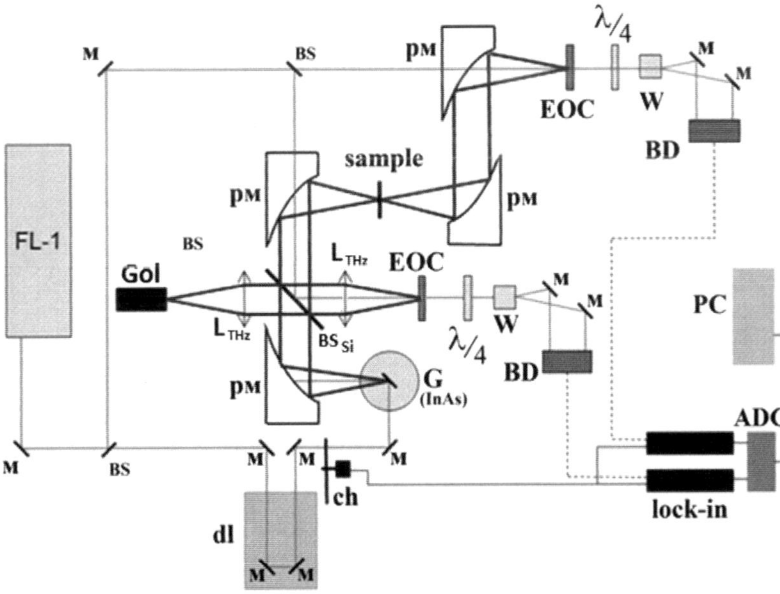

Figure 1. The scheme of general-purpose THz spectrometer. FL-1 - femtosecond laser; M - mirrors; BS - beam spleater; dl - optic delay line; G - THz radiation generator in InAs; pm - parabolic mirrors; ch - modulator; L_{THz} - lenses for THz radiation; EOC - electro-optical crystal CdTe; W - Wollaston prism; BD - balanced photodetector; lock-in - lock-in amplifier; ADC - analog-digital converter; PC - personal computer, Gol - Goley cell receiver.

Generated radiation passed through the teflon filter to cut off the wavelength range less than 50 microns, and then through the sample having a certain amplitude and phase transmittance. Passed THz radiation was focused on the EOC using two parabolic mirrors PM. Reflected from the object THz radiation is also collimated by an off-axis parabolic mirror PM, and after reflection from the beam splitter BS it was focused by THz lens LTHz with a focal length of 5 cm to electro-optical detector - CdTe crystal, that is cut in the direction of |100|. The sample was fixed perpendicularly to the optical axis in the focal plane using two axis motorized translation stage. EO detection was carried out using [10] quarter-wave plate, Wollaston prism, balanced photodetector and lock-in amplifier. Filtered and amplified signal was transmitted to a computer via a digital voltmeter. The magnitude of the error in determining the frequency during measurements was about 10 GHz [11]. Operation of the setup was realized by means of software environment NI LabVIEW.

When detecting terahertz radiation, we obtain the time dependence of the electric field E(t) of THz pulse at each point of the plane, so we can calculate the reference complex spectrum of the radiation incident on the sample $E_{ref}(\omega)$, by calculating the Fourier transform of the corresponding time sequence. By placing the required object in the path of THz radiation, it is possible to measure changed temporal form of THz pulse and complex spectrum of radiation transmitted through (reflected from) it $E_{obj}(\omega)$. The refractive index, the absorption coefficient and the dispersion of the investigated medium can be calculated using two obtained complex spectra [3, 6].

3. Experimental results

Using a laboratory setup (figure 1) reference THz pulses were obtained at the first stage of the researches. Air transmission for transmission mode (figure 2(a)) and the reflection from the golden mirror - for reflection mode (figure 2(b)) are presented. The generated THz pulse has single-period structure at the leading edge with a time width of the positive pulse peak at the half-amplitude about 1 ps. Oscillations with significantly smaller amplitude follow after the main single-period structure. They represent the dispersion properties of the medium. This is mainly due to the presence of water vapors in the air and their absorption lines in the THz frequency range.

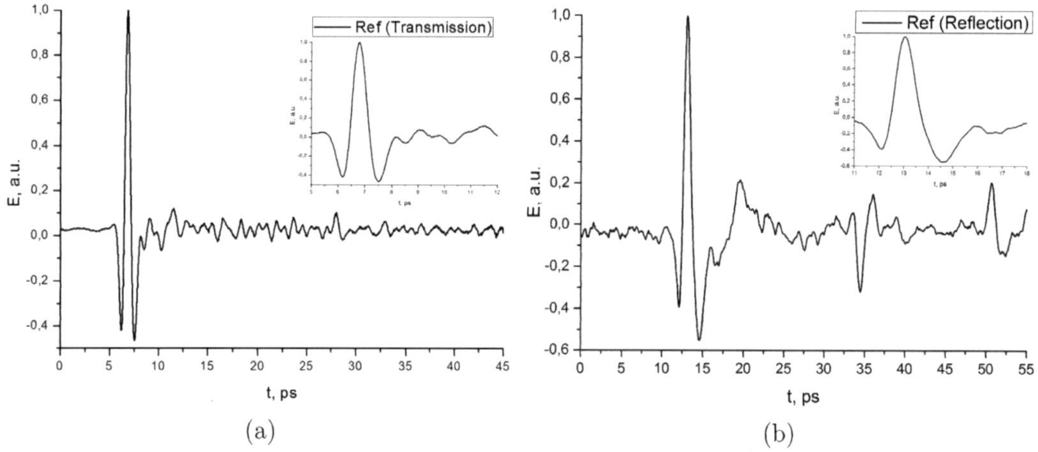

Figure 2. Dependence of THz pulse electric field intensity passed through air(a), reflected from gold mirror(b).

Single-period structure is observed in 34-36 ps (figure 2(b)), it corresponds exactly to the shape of the first one, but in antiphase and with a peak pulse amplitude about 0.3 from the first. This structure, which is delayed by 23 ps, is caused by the reflection of THz pulse from the second surface of the THz beamsplitter based on plate made of high-resistance silicon with a thickness of 1 mm (figure 1, BS_{Si}). The following temporary structures, in-phase with the first single-period structure, 27-29 ps (figure 2(a)) and 49-52 ps (figure 2(b)), correspond to the double reflection in CdTe crystal with a thickness of 3 mm. It is used as a detector of THz field (figure 1, EOC).

At the second stage transmission and reflection of THz radiation from the experimental objects for problems of biomedicine were investigated. Powdered amino acid Alanine, a human tooth and normal fibroblasts were used as objects. Dependence of THz pulse electric field intensity passed through and reflected from the samples is shown in figure 3 and figure 4.

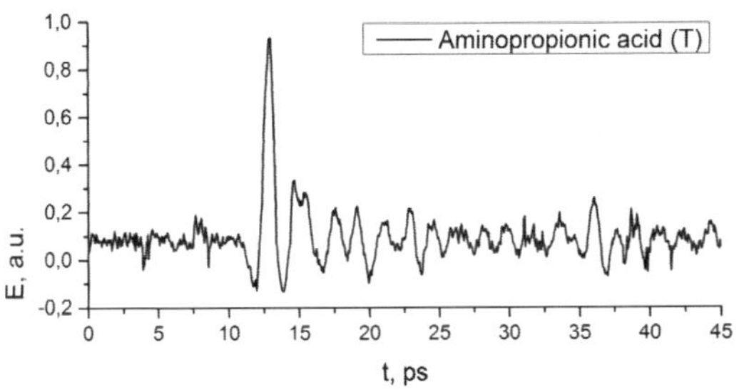

(a)

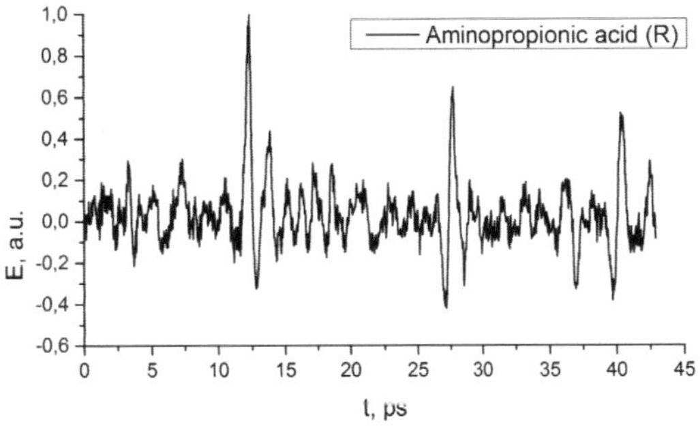

(b)

Figure 3. Dependence of THz pulse electric field intensity passed through amino acid (a), reflected from amino acid (b).

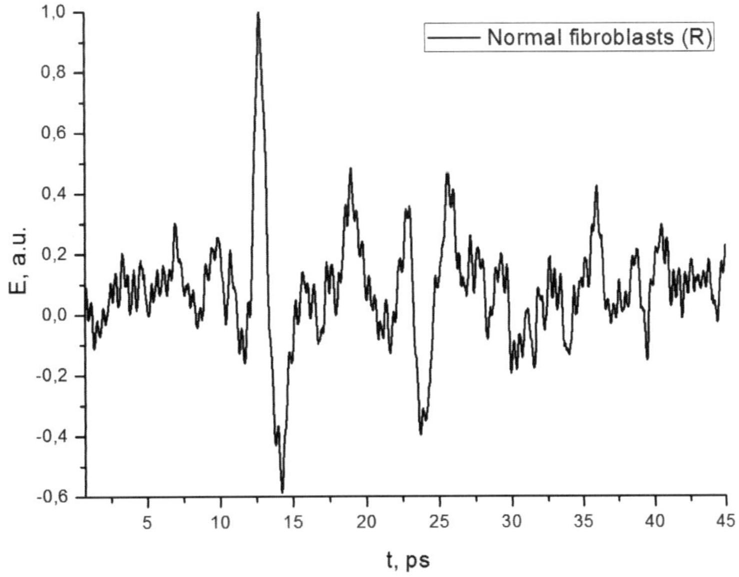

Figure 4. Dependence of THz pulse electric field intensity reflected from normal fibroblasts.

Experiments showed the opportunity to work with the powdered and liquid medias for carrying out their structural analysis without any extra preparations, tools and iterations. Using a laboratory setup experiments with the reflection of THz radiation on the human tooth were carried out.

In recent years, using of THz radiation for non-invasive determination of the state of teeth is rather perspective. Figure 5 shows tomographic image of human tooth obtained by raster scanning. It is clearly seen from the figure that the reflection from the front layer of enamel which has two interfaces - external layer and an inside layer. On the basis of graph, it is possible to determine the optical thickness of the enamel by measuring the time interval between pulse peaks. The reflection from dentin is observed on the back front of the reflected pulse, after reflection from the structure of tooth enamel.

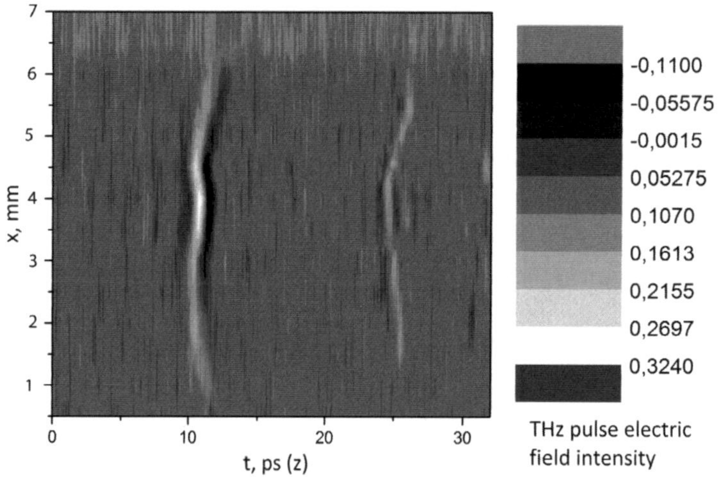

Figure 5. Tomographic image of human tooth obtained by raster scanning.

4. Conclusion

Laboratory setup is designed and developed for the measurements of transmission and reflection of THz radiation from a variety of media for biomedical problems. Dependences of THz pulse electric field intensity passed through / reflected from amino acid Alanine, dental tissue and normal fibroblasts were obtained using a laboratory setup. The possibility of studying the structure of powdered and liquid medias was demonstrated using this method.

Acknowledgments

This work was partially financially supported by Government of Russian Federation (Grant 074-U01). Nikolay Balbekin acknowledges financial support from Russian Ministry of Education and Science project within the state mission for institutions of higher education (Agreement No 2014/190).

References

[1] Auston D 1975 *Appl.Phys.Lett.* **26** 101-03
[2] Mourou G, Stancampiano C, Antonetti A and Orszag A 1981 *Appl. Phys. Lett.*. **39** 295-365
[3] Zhang X and Xu J 2009 (NY: Springer) 34-246
[4] Lee Yun-Shik 2007 (NY: Springer Science+Business Media) **LLC** 340
[5] Bespalov V, Gorodetsky A, Denysiuk I, Kozlov S, Krylov V, Lukomskii G, Petrov N and Putilin S 2008 *J. of Optical Technology* **75** 34-41
[6] Grischkowsky D, Keiding S, van-Exter M and Fattinger C 1990 *J. Opt. Society Am.* **7** 2006-15
[7] Gurtler A, Winnewisser C, Helm H and Jepsen P 2000 *J. Opt. Soc. Am.* **17** 74-83
[8] Izumida S, Ono S, Liu Z, Ohatake H and Sarukura N 1999 *Appl. Phys. Lett.* **72** 451-53
[9] Mittleman D, Cunningham J, Nuss M and Geva M 1997 *Appl. Phys. Lett.* **71** 16-18
[10] Lee Yun-Shik 2009 (NY: Springer) **XII** 340
[11] Chan W, Deibel J and Mittleman D 2007 *Prog. Phys.* **70** 1325-79

Journal of Physics: Conference Series **584** (2015) 012011 doi:10.1088/1742-6596/584/1/012011

Thin film thickness measurement error reduction by wavelength selection in spectrophotometry

Vladimir G Tsepulin*, Alexey V Perchik, Victor L Tolstoguzov and Valeriy E Karasik**

Bauman Moscow State Technical University, 2nd Baumanskaya str., 5, 105005, Moscow, Russia

E-mail: *tsepulin@gmail.com, **vitolstog@gmail.com

Abstract. Fast and accurate volumetric profilometry of thin film structures is an important problem in the electronic visual display industry. We propose to use spectrophotometry with a limited number of working wavelengths to achieve high-speed control and an approach to selecting the optimal working wavelengths to reduce the thickness measurement error. A simple expression for error estimation is presented and tested using a Monte Carlo simulation. The experimental setup is designed to confirm the stability of film thickness determination using a limited number of wavelengths.

1. Introduction

Thin film structures have many applications in industry. Current devices that contain multilayer thin film structures, such as liquid crystal and touch-screen displays and semiconductor devices, are mass-produced. Their quality often depends on the accuracy of the thin film parameters. The thickness is an essential property of a thin film that defines its physical characteristics. Therefore, a rapid yet precise technique for volumetric profilometry is needed to control the quality of film deposition during mass production.

A number of optical approaches are used to measure thin film thickness. Ellipsometry, spectrophotometry, and white-light interferometry are extensively applied for accurate and nondestructive film thickness measurement [1, 2, 3, 4, 5, 6, 7, 8, 9, 10, 11, 12]. All of these approaches allow us to measure not only the thickness of uniform thin films, but also the film thickness profile [1, 10, 5, 6, 7, 8]. Ellipsometry provides the most accurate thickness measurement, with errors as small as 5 Å [1]. Many measurements at different analyser positions are required to achieve this accuracy. In white-light scanning interferometry, the film thickness is computed by interferogram processing. Several intensity measurements at different optical path differences are needed to capture the required interferogram. Thus, spectrophotometry appears to be the most suitable approach for fast industrial profilometry.

Spectrophotometry is extensively applied for accurate and nondestructive film thickness measurement [3, 7, 5, 8, 4]. Several techniques have been developed for processing reflection spectra and determining the film thickness. A single-layer films thickness can be determined using a Fourier transform approach, but errors occur when the film thickness is less than 1 μm [13]. Nonlinear regression is a powerful approach that allows us to find the thickness of multilayer films or measure films that have unknown optical constants [3, 8]. The minimisation of the error function is usually a computationally intensive task, so nonlinear regression often

Content from this work may be used under the terms of the Creative Commons Attribution 3.0 licence. Any further distribution of this work must maintain attribution to the author(s) and the title of the work, journal citation and DOI.
Published under licence by IOP Publishing Ltd

Journal of Physics: Conference Series **584** (2015) 012011 doi:10.1088/1742-6596/584/1/012011

takes a considerable amount of time, although it can be accelerated by techniques such as direct phase calculation [6].

Quality monitoring systems must be able to measure the thickness of thin films with a lateral resolution of (10^6 points/cm^2). The speed of thickness determination becomes crucial in this task. One way to decrease the measurement and processing time is to reduce the amount of input information, but this also increases the thickness measurement errors.

When the number of wavelengths is limited, the measurement accuracy depends significantly on their values. The proper choice of wavelengths can considerably decrease the thickness determination errors.

Error analysis for multilayer film thickness determination by spectrophotometry is presented in [3]. The work in [14] presents an error analysis for a thickness determination technique based on the calculation of the optical parameters from three photometric quantities. The problem of choosing the wavelength values is not considered in the academic literature. In this paper, we introduce an approach to increasing the accuracy of thickness measurement by spectrophotometry using wavelength value selection. The optimal values are chosen using a priori information about the parameters of the film.

2. Theory

The reflection coefficient of a single-layer film can be expressed as follows:

$$R = |r|^2 = \left| \frac{r_{12} + r_{23} e^{2i\beta}}{1 + r_{12} r_{23} e^{2i\beta}} \right|^2, \tag{1}$$

where r_{12} and r_{23} are the Fresnel reflection coefficients for the airfilm and filmsubstrate interfaces, respectively; $\beta = \frac{2\pi}{\lambda} \hat{n}_2 d cos(\theta_2)$; $\hat{n}_2 = n_2 + ik_2$; d is the film thickness; θ_2 is the propagation angle of radiation in the thin film; n_2, k_2 are refractive index and extinction coefficient of the film, respectively.

Nonlinear regression is a common technique for processing reflection spectra. Assuming that the reflectivity of a thin film is distributed normally with a variance σ_i^2 for wavelength λ_i, the film thickness is the value that minimises the residual sum of squares:

$$\eta(d) = \sum_{i=1}^{N} \frac{1}{\sigma_i^2} \cdot [R_t(\lambda_i, d) - R_m(\lambda_i)]^2, \tag{2}$$

where $R_t(\lambda_i, d)$ and $R_m(\lambda_i)$ are the theoretical reflectance computed using (1) and the measured reflectance value, respectively.

3. Method to reduce thickness measurement error

The errors in film thickness determination by regression can be estimated using a linearised regression model. The standard deviation of the thickness can be expressed as [15]

$$\sigma_d(d_0, \lambda_1, \ldots, \lambda_N) = \frac{1}{\sqrt{\sum_{i=1}^{N} \left(\frac{1}{\sigma_i} \frac{\partial R}{\partial d}\Big|_{d_0,\lambda_i} \right)^2}}, \tag{3}$$

where $\frac{\partial R}{\partial d}\Big|_{d_0,\lambda_i}$ is the partial derivative value evaluated for thickness d_0 and wavelength λ_i.

The standard deviation depends on the derivative of the reflection, which depends on the thickness itself and the wavelength. Thus, the measurement error in a given thickness range can be decreased by selecting the optimal wavelengths.

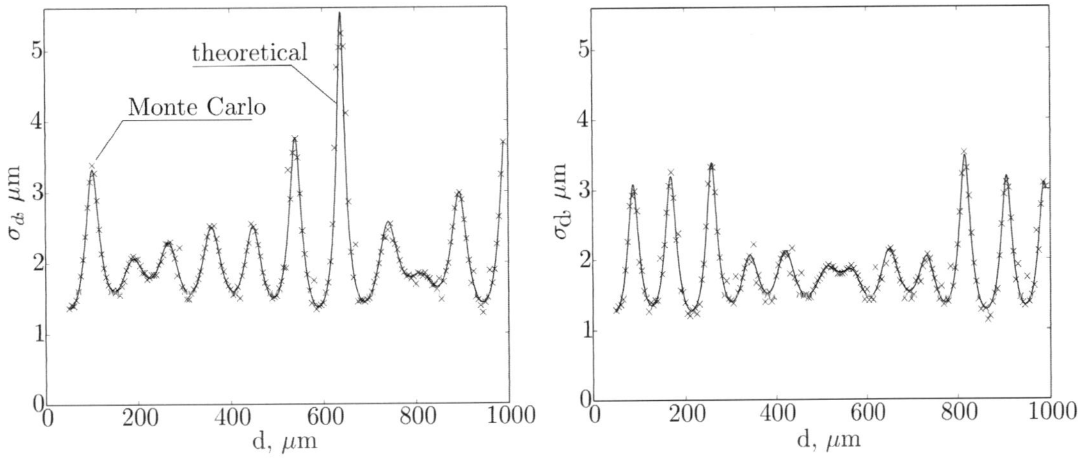

(a) $\lambda_1 = 500\ \mu m$, $\lambda_2 = 600\ \mu m$, $\lambda_3 = 700\ \mu m$ (b) $\lambda_1 = 456\ \mu m$, $\lambda_2 = 494\ \mu m$, $\lambda_3 = 741\ \mu m$

Figure 1. Standard deviation of film thickness measurement error dependence for thickness of MgF_2 film deposited on BK7 glass substrate, computed using expression (3) and Monte Carlo simulation.

For optimal wavelength selection, we introduce the objective function

$$Q(\lambda_1, \ldots, \lambda_N) = \min_{d \in [d_{min}, d_{max}]} \sum_{i=1}^{N} \left(\frac{1}{\sigma_i} \frac{\partial R}{\partial d}\bigg|_{d_0, \lambda_i} \right)^2, \tag{4}$$

which is the minimum value of the expression under the square root in the denominator of (3). As the function Q is, increases, the maximum thickness error in the given thickness range decreases. The values of λ_i, that maximise the function Q, are optimal.

Expression (3) was tested with Monte Carlo method [16]. For each thickness value, 2500 simulations were conducted. The reflectivity errors were sampled from the normal distribution $\mathcal{N}(0, \sigma)$. The film thickness was computed by minimising function (2) using the Levenberg-Marquardt minimisation algorithm [17]. The standard deviation computed from the resulting values as a function of the thickness for $N = 3$ is illustrated in figure 1, where N is the number of wavelengths. The simulation results show good agreement with the theoretical curve. The main standard deviation mismatch appears in the function extrema regions and is caused by nonlinearities in the function, which the employed model does not take into account.

To find the optimal wavelengths for measuring a MgF_2 film on BK7 glass substrate, function (4) was maximised using a differential evolution algorithm [18]. The thickness error for the obtained optimal wavelengths is shown in figure 1(b). The maximum standard deviation of the thickness error decreases from 5.24 μm for equally distributed wavelengths to 3.52 μm for the optimal wavelengths. The error value can be decreased by choosing a smaller thickness range.

The estimated error value for other numbers of wavelengths is shown in figure 2. The wavelength values were computed for each N value and a thickness range of 30 – 1000 nm using the proposed approach. The shaded area of the plot characterises all the possible values of the standard deviation at the determined wavelengths.

The measurement error decreases considerably with increasing N when the number of wavelengths is small (less than 10). Further increasing the number of wavelengths does not lead to such improvement.

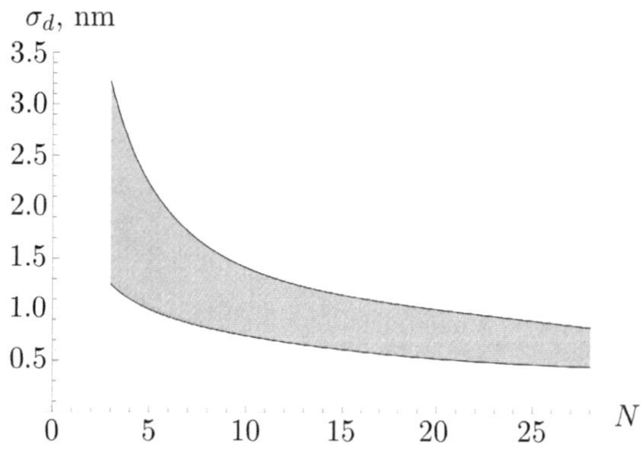

Figure 2. Standard deviation range for thickness measurement error as function of number of measured wavelengths. Wavelength values are optimised for thicknesses between 30 and 1000 nm.

Expression (3) can also be useful for estimating the error in multilayer film thickness measurement. The standard deviation of the k-th layer thickness can be found if the partial derivative is replaced by the corresponding $\frac{\partial R}{\partial d_k}$.

4. Experimental setup

Figure 3 shows schematic drawings of systems for film thickness measurement based on an acousto-optical tuneable filter (AOTF). A two-dimensional CCD sensor allows us to capture the spatial intensity distribution of light reflected from the sample, from which the spectral reflectances can be computed.

The setup shown in figure 3(b) has advantages over that in figure 3(a). The measurement error is proportional to the standard deviation of the noise [equation (3)], which can be improved by increasing the signal-to-noise ratio. The angle of light incidence can be adjusted to achieve the maximum image contrast and therefore the maximum signal-to-noise ratio. This angle equals the Brewster angle for a single-layer film deposited upon patterned structures.

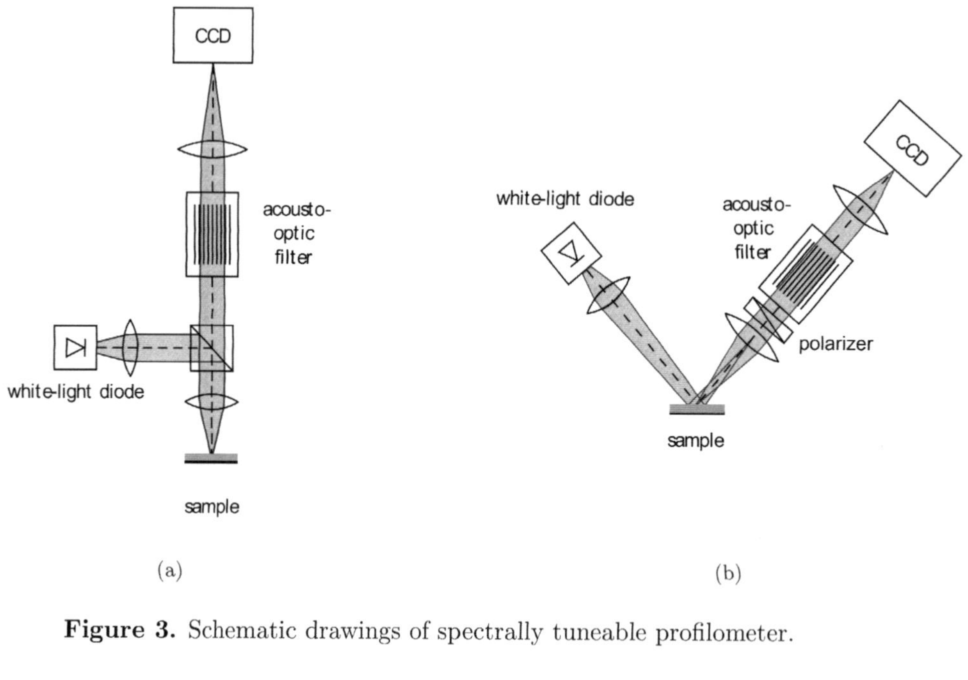

Figure 3. Schematic drawings of spectrally tuneable profilometer.

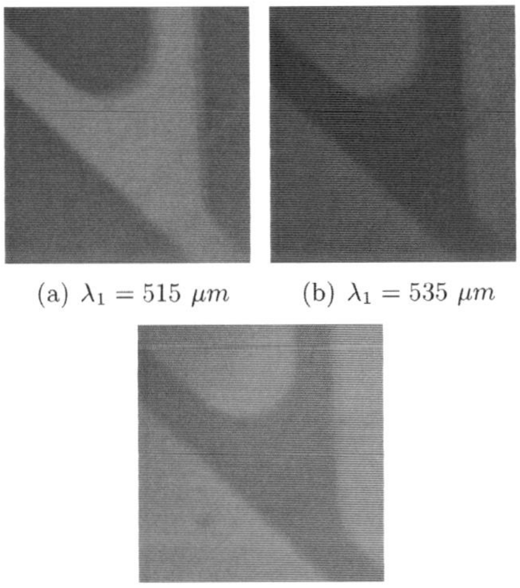

(a) $\lambda_1 = 515\ \mu m$ (b) $\lambda_1 = 535\ \mu m$

(c) $\lambda_1 = 585\ \mu m$

Figure 4. Spectral images of the sample.

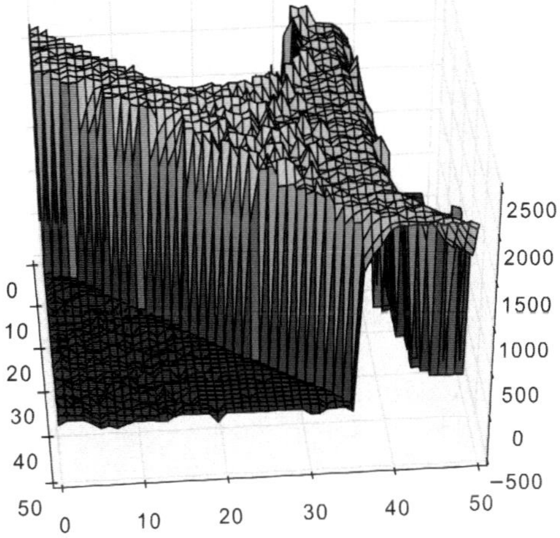

Figure 5. Computed thickness profile of the sample.

We also prefer the second setup because it does not contain a beam splitter, which causes considerable light intensity loss. The incidence angle can be tuned to achieve the best accuracy for single-layer and multilayer films deposited on substrates with various refractive indices.

To improve the spectral contrast, in our setup we used a double acousto-optical monochromator consisting of two non-collinear acousto-optical filters made of tellurium dioxide crystals and three polarisers (input, intermediate, and output) made of calcite crystals. The spectral contrast of a single AOTF is limited because of spectral side bands in its transmission function, which have the shape $\sin(x)/x$. These bands contain about 10%-20% of the energy, and using a double-AOTF design reduces that value to 0.3%-1%. In addition, using a double-AOTF design improves the image quality. The transmitted image quality is degraded by spectral drift of the image, the effect of different image scaling in two orthogonal directions, and nonlinear distortions. The use of a double-AOTF monochromator reportedly eliminates these effects because the second filter compensates for the image quality loss caused by the first AOTF [?].

The AOTF has the following characteristics:

- spectral range from 430 to 780 nm,
- 2 nm bandwidth,
- round input AOTF aperture with a diameter of 8 mm,
- 3° field of view,
- spatial resolution 800×800 elements.

Examples of spectral images are shown in figure 4. The thickness profile computed using the method described in section 2 is shown in figure 5. The experiment confirms that a unique solution can be found from three spectral reflectivity measurements.

5. Conclusion

Industrial systems for monitoring the quality of optoelectronic and semiconductor devices require techniques for thin film thickness measurement that are simultaneously fast and precise. We propose to use spectrophotometry-based systems with a limited number of wavelengths to address this problem. It is necessary to limit the input information for these systems to provide sufficient processing speed, but this also increase the measurement errors. Methods of reducing the errors in this case are not considered in other works.

In this paper, we introduce an approach to minimising the measurement error for a known film thickness range by proper wavelength selection. The proposed technique will be useful for industrial quality monitoring applications where the target thicknesses is usually known, and it is expected to considerably increase the monitoring accuracy (up to 40% compared with that obtained using arbitrarily chosen wavelengths).

We designed an experimental setup based on AOTF to test the applied algorithms. The processing of captured images confirmed the stability of the obtained solutions.

Acknowledgments

This work was supported by a grant from the Russian Foundation for Basic Research (RFBR 14-08-01103).

References

[1] Albersdörfer A, Elender G, Mathe G, Neumaier K, Paduschek P and Sackmann E 1998 *Appl. Phys. Lett.* **72** 2930–32
[2] Bader G, Ashrit P V and Truong V V 1998 *Appl. Opt.* **37** 1146–51
[3] Ylilammi M and Ranta-aho T 1993 *Thin Solid Films* **232** 56–62
[4] Horie M, Fujiwara N, Kokubo M and Kondo N 1994 Spectroscopic thin film thickness measurement system for semiconductor industries *Instrumentation and Measurement Technology Conference* vol 2 pp 677–82
[5] Kim D, Kim S, Kong H J and Lee Y 2002 *Opt. Lett.* **27** 1893–95
[6] Kim D and Kim S 2004 *Opt. Express* **12** 5117–24
[7] Qing D K, Yamaguchi I, Okamoto T and Yamamoto M 2000 *Opt. Lett.* **25** 914–16
[8] Wu K, Lee C C, Brock N J and Kimbrough B 2011 *Opt. Lett.* **36** 3269–71
[9] Ghim Y S, Suratkar A and Davies A 2010 *Opt. Express* **18** 6522–29
[10] Kim S W and Kim G H 1999 *Appl. Opt.* **38** 5968–73
[11] Hlubina P, Ciprian D, Luek J and Lesk M 2006 *Opt. Express* **14** 7678–85
[12] Kassamakov I, Hanhijärvi K, Abbadi I, Aaltonen J, Ludvigsen H and Hæggström E 2009 *Opt. Lett.* **34** 1582–84
[13] Merklein T M 1990 *Appl. Opt.* **29** 505–11
[14] Konstantinov I, Babeva T and Kitova S 1998 *Appl. Opt.* **37** 4260–67
[15] Bates D M and Watts D G 1988 *Nonlinear Regression Analysis and Its Applications* (John Wiley & Sons, Inc.)
[16] Metropolis N and Ulam S 1949 *J. Am. Statist. Assoc.* **44** 335–41
[17] Marquardt D W 1963 *J. Soc. Ind. Appl. Math.* **11** 431–41
[18] Storn R and Price K 1997 *J. Global Optim.* **11** 341–59
[19] Pustovoit V I, Pozhar V E, Mazur M M, Shorin V N, Kutuza I B and Perchik A V 2005 Double-aotf spectral imaging system *Acousto-optics and Photoacoustics* vol 5953

Influence of the planar cylindrical target erosion zone of magnetron sputtering on the uniformity of a thin-film coating

D D Vasilev and K M Moiseev
Bauman Moscow State Technical University, 2nd Baumanskaya str., 5, 105005, Moscow, Russian Federation

E-mail: d.d.vasiliev@ya.ru, mkm430@ya.ru

Abstract. The constant change in the erosion zone of magnetron sputtering targets makes simulating their profile difficult, whilst the measurement of real-world systems requires a long working time. However, by using the measured erosion zone profile of a real target, a formula for calculating the uniformity of thin films has been developed. Using this, it has been found that the erosion zone of a 4” cylindrical magnetron target can improve coating uniformity.

1. Introduction

According to Mur’s law, the complexity and size of semiconductor crystals double every year, yet the dimensions of semiconductor devices are halved. Over the last 15 years this has seen the size of the smallest unit decrease ten-fold, but with a doubling of the diameter of their substrate seven-fold increase in the dimensions of the chips in which they are used. This microminiaturization makes it far more important to create a high degree of uniformity in the layer thickness through careful monitoring of the fabrication process and imposing much stricter requirements to ensure a clean technological environment [1]. For example, a past increase in substrate diameter from 200 to 300 mm resulted in the layer uniformity changing from ±5% to ±4%, with current technology now being designed around 450 mm diameter substrates.

Of the newer microelectronic technology, single infrared photon detectors have been receiving particular attention of late, and are based on a 4.6 nm thick superconducting WSi film deposited by DC magnetron co-sputtering from separate W and Si targets at room temperature [2]. The distribution of material during magnetron sputtering, however, can lead to variation in material ratio at different points on the substrate that can cause the film to lose its superconductivity. Various forms of substrate holder rotation have been tested in a bid to alleviate this problem, but the design and experimental testing of new equipment takes enormous time and financial resources.

Computer simulation represents a way of reducing the cost outlay of improving the film uniformity of magnetron sputtering, but a key requirement of this is an ability to take into account the constantly changing shape of the magnetron target erosion area. However, registration of this erosion area is rather time-consuming due to the fact that it is unique for each magnetron sputtering source, magnetic system, target material, type of power supply, etc. The purpose of this article is therefore to provide a theoretical evaluation through mathematical modelling of the influence of erosion area on the uniformity of thin film layer produced by planar cylindrical magnetron sputtering source.

Content from this work may be used under the terms of the Creative Commons Attribution 3.0 licence. Any further distribution of this work must maintain attribution to the author(s) and the title of the work, journal citation and DOI.
Published under licence by IOP Publishing Ltd

2. Physical phenomenon, underlying process

In magnetron sputtering systems for diode-type devices, sputtering occurs through bombarding a target surface with gas ions formed through an abnormal glow discharge plasma. As seen in figure 1, cavities for water cooling and magnets to confine the plasma are installed in the housing of the source, with the target of sputtering being fixed from above. The gas ions require a critical energy before sputtering can occur, with the number of atoms ejected by a single ion (the sputtering coefficient) being dependent on not just the ion's energy, but also the tilt angle and material used.

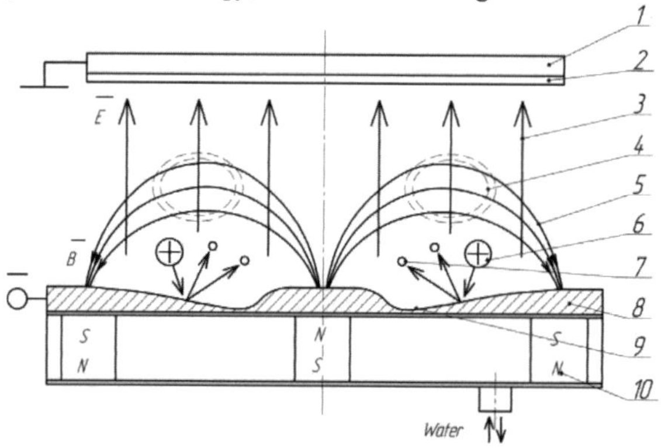

Figure 1. Schematic diagram of a magnetron sputtering system with a planar target: 1 – substrate holder; 2 – substrate; 3 – electric field; 4 – area of plasma localization; 5 – lines of magnetic force; 6 – ions; 7 – sputtering fractions; 8 – target; 9 – area of erosion; 10 – magnets.

The shape of the target surface (figure 2) can be explained by the non-uniformity of the plasma localization above it, which in turn is due to inconsistency in the ion current density. The erosion area of the target is therefore directly proportional to the ion current density, with the lowest point in its topography coinciding with the highest sputtering rate, whereas at the highest points the rate is zero. This means that in effect the sputtering rate can be considered directly proportional to the topography of the erosion area.

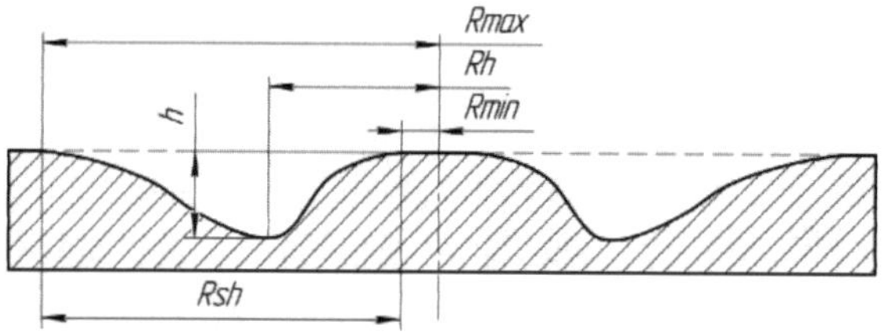

Figure 2. Erosion area of a magnetron sputtering source target: Rmax – peak radius; Rmin – minimum radius; Rh – distance to the lowest point; Rsh – width of erosion area; h – depth of erosion area.

The distribution of magnetron sputtered material occurs in accordance with the law of cosines to a degree of *n* [3], a coefficient which is determined by:

- The material, uniformity, grain dimensions and structure of the target, which represent the most significant factors [4].
- The influence of electric and magnetic fields on the trajectory of the sputtered material's ionized atoms.
- The collision of sputtered atoms with each other and with the working gas.
- An absence of sputter atom condensation at the point of impact with the substrate
- Variation in the target material and the density of the deposited film.

Experimental calculations of n for various targets, magnetron sputtering sources and modes of deposition suggest a typical range of between 0.4 and 1.7 [3]; however, computer simulation can take into account a greater of parameters [5, 6].

3. Mathematical description

The thickness of a film deposited by source evaporation from a point source can be determined for any given location on the substrate δ (figure 3) by:

$$h = \frac{dq_u(\varphi,\theta)}{\rho dA_0} = \frac{q_u \cos(\varphi)\cos(\theta)}{\pi C^2 \rho} \text{m},$$

where ρ is the thickness of the material, q_u is the weight of sputtered or evaporated material in kg, ϕ is the angle between the direction of evaporation and the selected point on the substrate, C is the distance and θ is the angle of deposition material.

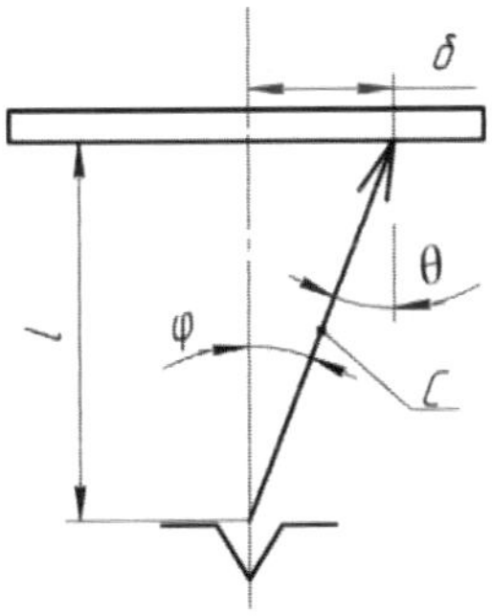

Figure 3. Geometric scheme for calculating the thickness of a film produced by magnetron sputtering from a point source.

As the process of magnetron sputtering occurs in the erosion area, the uniformity of the film needs to be calculated from the various different point sources that exist within this area. This is because each point has its own sputtering rate, which is directly proportional to the erosion area's topography. This topography is determined mostly by plasma localization above the surface of the source; and consequently, depends on the magnetic field. Calculation of the topography based on measurements of the magnetic field therefore allows for a prediction error of 15 % for balanced magnetron sputtering sources and 25 % for unbalanced magnetron sputtering sources [9]. To reduce this error, the topography of the erosion area has instead been measured directly; a method which has had some success by allowing individual points to be connected by cubic splines (figure 4).

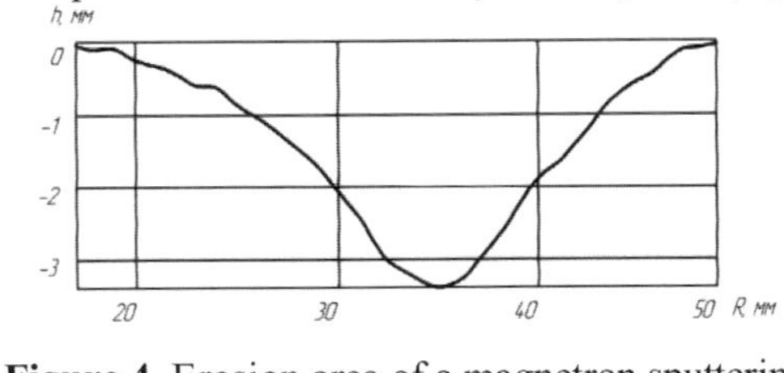

Figure 4. Erosion area of a magnetron sputtering source.

Owing to the fact that calculation of the degree of non-uniformity in the film thickness relates directly to the maximum thickness, the constants in the formula can be omitted. The magnetron sputtering shown in figure 5 can therefore be described mathematically by:

$$T(r,t) = \int\limits_{\alpha}\int\limits_{R} \frac{\cos^n(\varphi(R,r,\alpha,t)) \cdot \cos(\theta(R,r,\alpha,t)) \cdot \sqrt{1+(f_R^{'}(R,t))^2} \cdot g(R,t) \cdot R}{C(R,r,\alpha,t)^2} dR d\alpha$$

where r is the distance to the point at which non-uniformity is calculated, t is the time, R is the radius of the erosion area (which changes from $Rmin$ to $Rmax$), α is the angle of integration in the cylindrical coordinates (which changes from 0 to 2π), n is the coefficient for the target magnetron sputtering source and process parameters, ϕ is the angle of particle motion relative to the normal of the erosion area surface, θ is the angle of particle motion relative to the normal to the substrate surface, $f(R)$ is a function of the erosion area topography, $g(R)$ is a sputtering rate function, and C is the distance between the sputtering point and deposition zone.

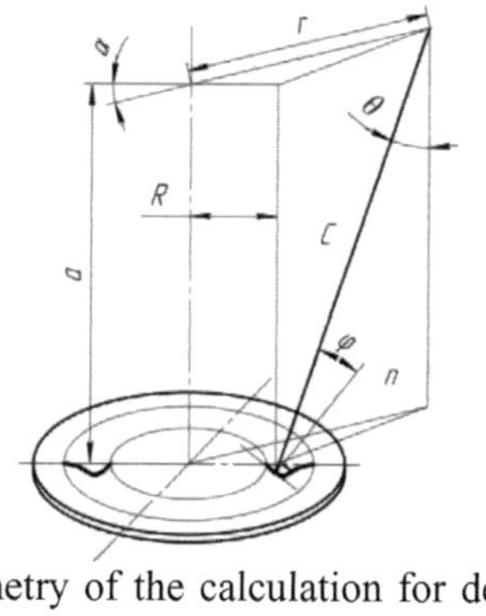

Figure 5. The geometry of the calculation for determining the non-uniformity in a coating produced by magnetron deposition.

This formula for determining the non-uniformity of a coating can be determined as:

$$P(r,t) = \frac{T(r,t)}{T(0,t)}$$

Note that this simplification assumes that the degree of reverse sputtering is zero, and is only applicable in instances where the entire substrate is in the area of erosion.

To determine the cosine degree, it is necessary to first carry out a series of experiments under a given set of parameters to determine the thickness of the film deposited, with subsequent calculation being complicated by its dependence on many factors. Determining the topography of the sputter target requires experiments over long periods of time, creating a need to instead estimate how the erosion area affects the uniformity of coating.

4. Computer simulation

The magnetron sputtering model developed was designed to approximate the vacuum coating system VUP-11M located at the Electronic Technology in Engineering department of BMSTU. The diameter of the magnetron is 100 mm, and a measured area of its target erosion was used in the simulation (figure 4). All simulation was carried out using the program MathCad, with calculations being performed based on a substrate to target distance of 100 mm (figure 6).

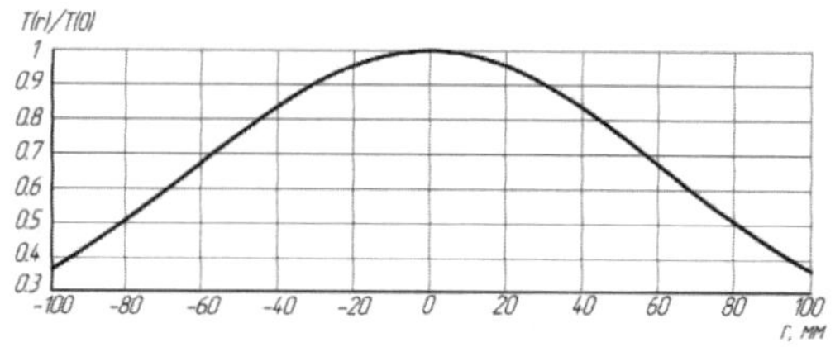

Figure 6. Graph of the uniform coating generated by a 100 mm diameter magnetron sputtering source at a distance of 100 mm from the target.

Simulation was conducted based on equation 1 of the coating uniformity achieved by sputtering a target with an erosion area (figure 4), without an erosion zone (i.e., a new target), and without an erosion area but taking into account the sputtering rate. The results obtained (Table 1) show that without an erosion area and registration of the rate distribution *g*, the non-uniformity of the coating is increased (0.4 % at a distance of 50 mm, 0.7 % at 100 mm, 0.5 % at 150 mm.). Meanwhile, increasing the minimum depth of the erosion area from 0 reduces this non-uniformity, which means that a target without erosion will produce a far less uniform coating than a target that is eroded.

Table 1. Simulation of the coating non-uniformity generated by magnetron sputtering (magnetron diameter of 100 mm, distance of 100 mm).

Distance from the target's centre	With erosion area and sputtering rate distribution accounting	Without erosion area (new target) but with sputtering rate distribution accounting	With erosion area, but without sputtering rate distribution accounting	Without erosion zone and sputtering rate distribution accounting
50 mm	24,2%	24,7%	24,3%	24,6%
100 mm	63,3%	64,1%	63,5%	64,0%
150 mm	84,5%	85,0%	84,6%	85,0%

5. Conclusions

On the basis of the computer simulation results presented it can be said that increasing the erosion area has a negligible (<1%), but nevertheless positive, effect on the uniformity of a magnetron sputtered coating. This, of course, assumes that the entire substrate is within the area of erosion. The development of deposition parameters for obtaining a desired uniformity therefore needs to consider not just the erosion area and the sputtering rate distribution, but also the fact that the increase in erosion area over time will improve the coating's uniformity.

References

[1] Chstiakov U and Rainova U 2010 *Introduction to the processes of integrated micro-and nanotechnology* (Moscow: BINOM) p 10
[2] Verma V B, Horansky R, Marsili F, Stern J A, Shaw M D, Lita A E, Mirin R P and Nam S W 2014 *J. Appl. Phys.* **104** 051115
[3] Martinenko U, Rogov A and Shulga V 2012 *Tech. Phys.* **57** 439
[4] Akulenok M, Andreev V and Gromov D 2011 *Introduction to the processes of integrated micro-and nanotechnology* (Moscow: BINOM) p 56
[5] Golosov D, Melnikov S and Dostanko A 2012 *Surf. Eng. Appl. Electrochem.* **48** 52
[6] Huang Y, Gao S T and Liu M 2011 *Advanced Engineering Forum* **2-3** 1082

Wavelet-domain de-noising of optical coherent tomography data for biomedical applications

Kirill I Zaytsev[1]*, Konstantin G Kudrin[2], Nikita V Chernomyrdin[1], Alexey M Khorokhorov[1], Alexander B Prytov[1], Irina N Dolganova[1], Alexey V Perchik[1], Igor V Reshetov[3] and Stanislav O Yurchenko[1]**

[1]Bauman Moscow State Technical University, 105005, 2nd Baumanskaya str. 5, Moscow, Russia;
[2]P.A.Hertzen Moscow Research Institute of Oncology, 125284, 2nd Botkinskiy proezd 3, Moscow, Russia;
[3]Institute of Improvement of Professional Skill of the Federal Medico-Biological Agency of Russia, 125371, Volokolamskoye shosse 91, Moscow, Russia

E-mail: `*kirzay@gmail.com, **st.yurchenko@mail.ru`

Abstract. Optical coherent tomography (OCT) is a rapidly developing method of fundamental and applied research. Detection and processing of OCT images is a very important problem of applied physics and optical signal processing. In the present paper we are demonstrating the ability for effective wavelet-domain de-noising of OCT images. We are realizing an algorithm for wavelet-domain de-noising of OCT data and implementing it for the purpose of studying test samples and for *in vivo* nail tomography. High de-noising efficiency with no significant losses of information about the internal sample structure is observed.

1. Introduction and background

Optical coherence tomography (OCT) is an imaging technique allowing non-invasive *in vivo* and *in vitro* three-dimensional imaging of both transparent and nontransparent biological tissues on a micrometer scale with no necessity of contact between the probe and the tissue [1–14].

OCT produces an image of optical scattering from internal tissue microstructures by measuring the echo time delay and magnitude of backreflected light in a way that is analogous to ultrasonic pulse-echo imaging [15] or imaging the object with terahertz time-of-flight tomography [16–19]. The result of OCT imaging is a profile of backreflection intensity, containing information about the internal structure of the sample. Because the high speed of light makes a direct measurement of optical echo impossible, OCT uses interferometric techniques with ultrashort light pulses or low coherent light sources to range distances at the level of single micrometers [2-14]. OCT was first developed by Fujimoto's group [1], and the method has since matured into an important clinical imaging modality. Since OCT is non-invasive technique, it becomes widely applicable in medical disciplines [20] such as ophthalmology [21–24] gastroenterology [25–28] dentistry [29–32] and dermatology [33–36]. An important aim of OCT is to acquire morphological changes within biological tissue. Due to its capability to study the structure of inhomogeneous objects, ability to work in strongly dispersive media, high potential in obtaining cellular level resolution, and non-ionizing nature of radiation presence, OCT appears to be the most promising method of examination and characterization of epithelial tissue *in vivo*.

Content from this work may be used under the terms of the Creative Commons Attribution 3.0 licence. Any further distribution of this work must maintain attribution to the author(s) and the title of the work, journal citation and DOI.
Published under licence by IOP Publishing Ltd

This paper presents the algorithm for effective de-noising of OCT data in wavelet domain. We are developing an algorithm for wavelet-domain de-noising of OCT signals and implementing it for studying the test samples, as well as for tomographic reconstruction of nail structure *in vivo*. The results of algorithm implementation and possible applications of the proposed technique are discussed.

2. Materials and methods

At first, we are briefly discussing the techniques of time-domain and Fourier-domain OCTs [3]. The principle of OCT operation is based on the low-coherence interferometry, and the key component of every OCT system is a two-beam interferometer (typically Michelson's interferometer) with low-coherence and broad-bandwidth light source (figure 1, 2). The light beam generated by the source is divided into two parts at the beam splitter. The first one is propagating towards the reference mirror, and the second one is directed to the object of interest. After the waves were scattered on the surface of reference mirror and on the surface of the object, the scattered light components travel to the detector. Let us define the optical time-delay of the reference beam with duration t_r, and the sample beam time-delay with t_s. The interference pattern occurs on the surface of the detector as a result of coherent superposition of the wave amplitudes, but interference could appear only for time-delays of optical beams satisfying the condition

$$|t_s - t_r| \leq \frac{t_c}{2} = c\frac{l_c}{2} \tag{1}$$

where t_c and l_c are coherence time and coherence length, respectively, and $c \approx 3 \cdot 10^8$ m/c is the speed of light in vacuum. By time-gating the echo response, we are able to create a depth-resolved line profile of the tissue (or A-scan). An OCT image (B-scan) is built up as a series of adjacent axial depth scans (A-scans). In OCT the depth-resolved information about internal structure of the sample can be obtained from either the time-domain measurements (TD-OCT) or the Fourier-domain (FD-OCT) measurements [3].

Figure 1 shows a schematic representation of TD-OCT set-up and describes general principles of measuring the profile of backscattering intensity, $I(z)$ [3]. The main advantage of TD-OCT is an automatic removal of coherence noises caused by interference of multiple beams reflected from various components of the interferometer optical systems, and this is due to the time-gating principle of TD-OCT operation. The principle of FD-OCT operation are based on the Wiener-Khintchine theorem [45]. It approves that the profile of backscattered light intensity

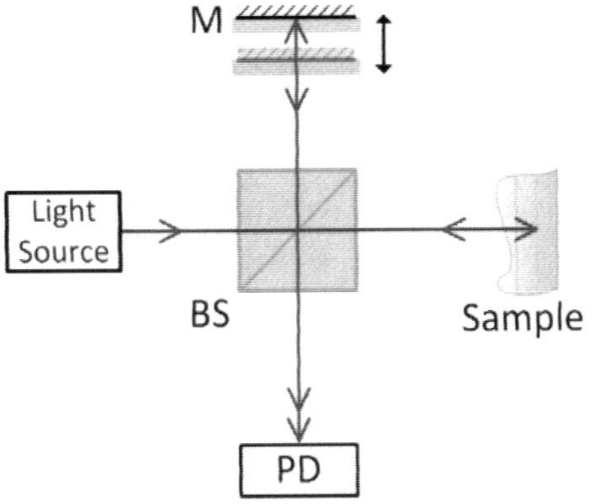

Figure 1. Schematic representation of TD-OCT measurements: M stands for the reference mirror of Michelson's interferometer, BS stands for the beamsplitter, and PD stands for the photodetector. In TD-OCT the backscattering intensity profile $I(z)$ is detected by means of post-processing the series of interference patterns detected at PD while continuous translating M. Since the low-coherence light source is utilized for pumping the sample, each position of M corresponds to the certain optical depth of light backscattering. Coherence length l_c defines the theoretical limit of TD-OCT depth-resolution as $\Delta z \geq l_c$.

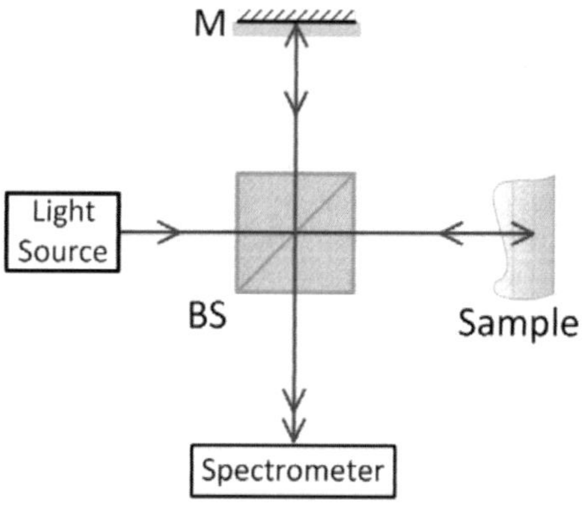

Figure 2. Schematic representation of FD-OCT measurements: M stands for the reference mirror of Michelson's interferometer, and BS stands for the beamsplitter. In FD-OCT the spectral dependence of sample complex amplitude reflectivity $\widetilde{R}(\nu)$ is detected by means of post-processing the interferometric data. The interference pattern is being detected for a different light frequencies ν either by tuning the monochromatic wavelength of light source or by simultaneous detection of interference patterns at different wavelengths. Reconstruction of the internal structure of the sample is based on complex reflectivity $\widetilde{R}(\nu)$ and the Eq. (2). The bandwidth of sample reflectivity registration $\Delta\nu$ limits the depth resolution of FD-OCT measurements, $\Delta z \geq 1/\Delta\nu$.

$I(z)$ is directly related to the spectral dependence of the sample amplitude reflectivity, and the latter could be reconstructed after interferometric measurements. Let us define the amplitude reflectivity of sample with $\widetilde{R}(\nu)$, where ν is a frequency of electromagnetic wave, then the backscattering intensity profile could be reconstructed in the following way

$$I(z) \propto |R(t)|^2 = \left|\mathcal{F}_{\nu}^{-1}\left[\widetilde{R}(\nu)\right]\right|^2 \tag{2}$$

where $R(t)$ corresponds to the sample pulse response, and $\mathcal{F}_{\nu}^{-1}[...]$ stands for the inverse Fourier transform. Figure 2 shows the principles of FD-OCT operation.

Two types of FD-OCT exist: 1) spectral-domain OCT captures the sample reflectivity simultaneously in the entire spectral range of the system sensitivity by using the spectrometer equipped with a high-speed detector; 2) swept source OCT captures the spectral dependence of the sample amplitude reflectiviy with a single detector by means of sweeping the output frequency of tunable laser source. The main advantage of FD-OCT is associated with the ability for fast detection of backscattering intensity profile without any mechanical scanning, and the acquisition speed is only limited with the detector frame rate. Modern OCT systems with enhanced axial resolution of 3 μ allows obtaining *in vivo* backscattering intensity profiles. In clinical practice, the ability for visualizing the subtle morphological features of tissue significantly improves the early diagnosis and the assessment of tissue pathologies.

3. Wavelet-domain de-noising approach

Modern OCT systems, based on various operational principles, allow accurate characterization of *in vivo* biological tissue [1-36], however, the problem of OCT signal post-processing, including solution of the backscattering inverse problem and signal de-noising, still remains very important for further development of OCT applications. We are demonstrating the ability for highly-efficient de-noising of OCT data by means of wavelet-domain prossing [37–39]. The techniques for wavelet-domain de-noising seem to be very effective for processing of optical signals [40–41], including de-noising of spectroscopy data [42–44] due to high locality of wavelet transform kernels both in time-domain and frequency-domain. We are experimentally implementing wavelet-domain de-nosing for processing the OCT measurements.

Journal of Physics: Conference Series **584** (2015) 012013 doi:10.1088/1742-6596/584/1/012013

Let us define $I(x,z)$ to be the results of OCT measurements, where z stands for the sample depth, and x stands for the lateral scanning direction. Thus, the signal $I(x,z)$ is the result of measuring of the backscattering intensity profile $I(z)$ at different points x of sample surface. We are applying the wavelet-domain de-noising separately for various lateral coordinates, thus, let us consider the procedure for single backscattering intensity profile $I(z)$ corresponding to the certain $x = x'$.

The wavelet-domain de-nosing procedure is used in the following way. At first, the wavelet-domain decomposition is applied to the backscattering intensity profile

$$C(a,b) = \mathcal{W}\left[I\left(z\right) - m.e.\right] = \int_{-\infty}^{+\infty} \left(I\left(z\right) - m.e.\right)\psi\left(a,b,z\right)dz, \tag{3}$$

where $m.e. = \frac{1}{Z}\int I(z)dz$ is a mathematical expectation of the intensity profile (Z is an integration interval), $\mathcal{W}[...]$ is a direct wavelet transform operator, $C(a,b)$ are wavelet decomposition coefficients, and $\psi(a,b,z)$ are wavelet decomposition kernels. Note, all $\psi(a,b,z)$-kernels are constructed by scaling and shifting of the mother wavelet, $\psi(z)$, as $\psi(a,b,z) = |a|^{-1/2}\psi\left(\frac{z-b}{a}\right)$ (a defines the kernel scale and b defines the kernel translation).

De-noising is implemented via the thresholding of the wavelet decomposition coefficients

$$C_T\left(a,b\right) = \begin{cases} C\left(a,b\right), & if \quad C\left(a,b\right) \geq T, \\ 0, & if \quad C\left(a,b\right) < T, \end{cases} \tag{4}$$

where $T = \sqrt{2\ln(n)}$ defines the universal threshold, utilized in the present work, and n stands for the number of wavelet-decomposition coefficients. Inverse continuous wavelet transform is applied to de-noised wavelet spectrum and the mathematical expectation is being added

$$I_T\left(z\right) = \mathcal{W}^{-1}[C_T\left(a,b\right)] + m.e. = C_\psi^{-1}\iint_{-\infty}^{+\infty}\frac{C(a,b)}{a^2}\widetilde{\psi}\left(a,b,z\right)dadb + m.e., \tag{5}$$

where $\mathcal{W}^{-1}[...]$ is an inverse wavelet transform operator, $\widetilde{\psi}(a,b,z)$ is the dual function of $\psi(a,b,z)$, and

$$C_\psi = \int_{-\infty}^{+\infty}\frac{\Psi(\omega)\widetilde{\Psi}(\omega)}{|\omega|}d\omega < \infty \tag{6}$$

is the admissible constant. Functions $\Psi(\omega)$ and $\widetilde{\Psi}(\omega)$ in (6) correspond to Fourier spectra of $\psi(z)$ and $\widetilde{\psi}(z)$, respectively; and the admissible constant (6) restricts the diversity of functions suitable for definition of $\psi(z)$.

We use fast wavelet-transform (FWT) algorithm with *'coif5'* mother wavelet, which was found to be optimal for the purpose of OCT data decomposition [42–44], and FWT helps to implement real-time data processing.

4. Results and discussion

Let us describe the results of the wavelet-domain de-noising implementation for processing the test OCT signals. We utilize TD-OCT set-up with fiber probe for this purpose. The set-up is based on picosecond Er-doped fiber laser and fiber-based Michelson's interferometer equipped with mechanical scanning system.

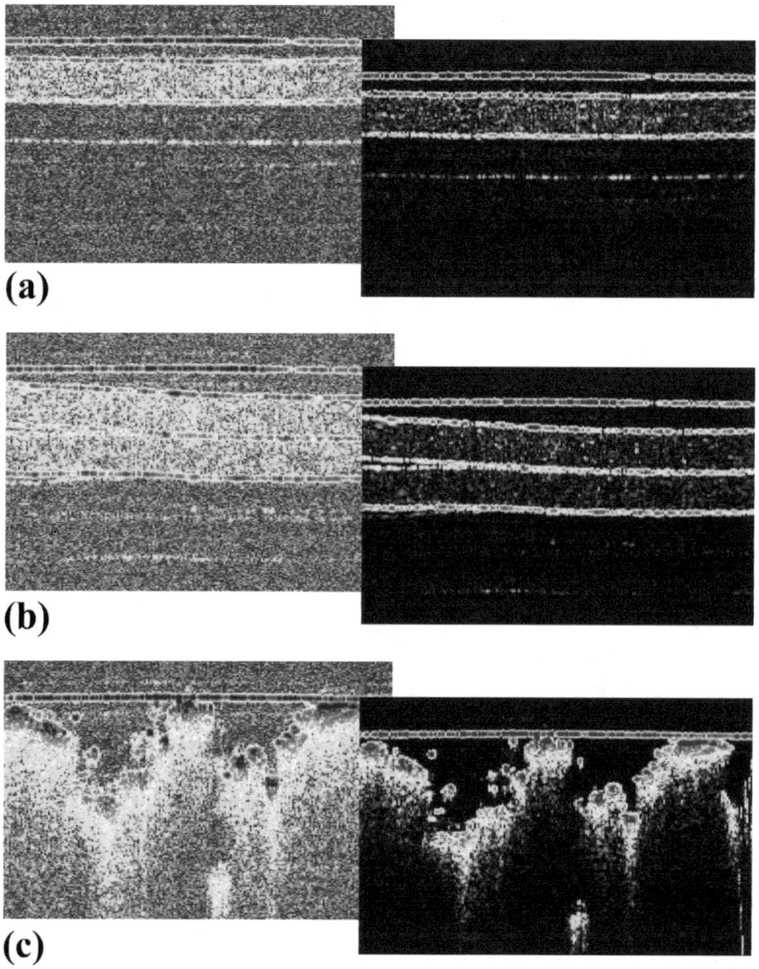

Figure 3. Results of wavelet-domain de-noising of time-domain OCT data: (**a**) corresponds to the test sample made of single-layer polymer film, (**b**) corresponds to the test sample made of double-layer polymer film, and (**c**) corresponds to the test sample made of bint. The images of left and right columns show initial noisy tomographic images and tomographic images after de-noising, respectively.

Figure 3 shows the results of studying the test samples of polymer films and complex scattering sample made of bint. One should notice significant noise suppression for all of the test signals. The contrast in images has been raised dramatically with no losses of important information about the sample structure. Note, the top response in all tomographic images corresponds to the light backscattering on the reference plane window, which is utilized in order to possess OCT probe on the surface of the studied object. We could see strong backscattering from all the interfaces of single-layer and two-layer polymer film. Moreover, the backscattering on the internal inhomogeneities of the film is still notable after the noise suppressing. The noises in bint image are also suppressed; and even small scatterers are notable in the tomographic image.

We have studied *in vivo* samples of human nail as a representative example of biomedical

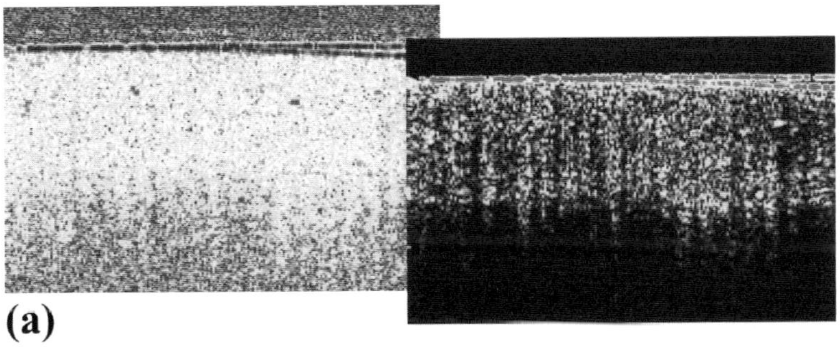

(a)

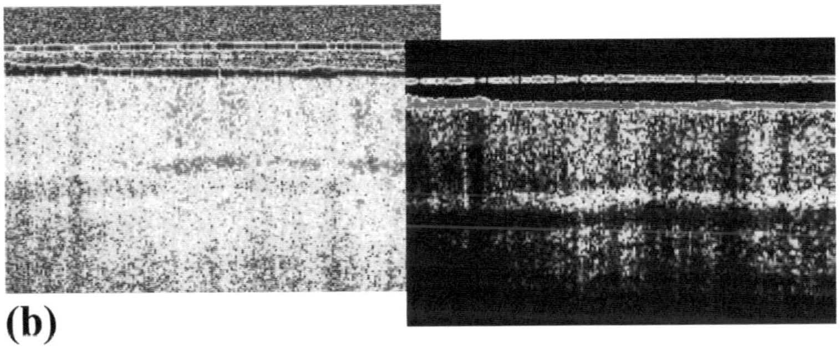

(b)

Figure 4. Results of wavelet-domain de-noising of time-domain OCT data: **(a)** corresponds to the nail body tomographic image and **(b)** corresponds to the tomographic image of nail lunula. The images of Left and right columns show initial noisy tomographic images and tomographic images after de-noising, respectively.

algorithm implementation. Tomographic images of nail body and nail lunula have been considered. We have significantly suppressed the noise in nail images, and all the structural features of tissue are notable in the tomogram. Backscattering on small inhomogeneities of nail is present in the image of nail body both before and after data de-noising. Image of nail lunula contains the responses from several interfaces between different tissues in a sample, allowing us to analyze the internal structure of nail and underlying tissue.

5. Conclusions

In the present paper we demonstrated an ability for effective wavelet-domain de-noising of OCT data. We have realized an algorithm for TD-OCT data de-noising in wavelet-domain and implement it for studying the test samples. High de-noising efficiency with no significant losses of information about the object structure have been demonstrated.

Acknowledgements

This work was supported by the Russian Scientific Foundation, Project # 14-15-00758.

References

[1] Huang D, Swanson E A, Lin C P, Schuman J S, Stinson W G, Chang W, Hee M R, Flotte T, Gregory K, Puliafito C A, and Fujimoto J G 1991 *Science* **254** 1178–81
[2] Wojtkowski M 2010 *Applied Optics* **49** D30–D61
[3] Leitgeb R A, Hitzenberger C K, and Fercher A F 2003 *Opt. Exp.* **11** 889–94
[4] Leitgeb R A, Hitzenberger C K, Bajraszewski T, and Fercher A F 2003 *Opt. Lett.* **28** 2201–04
[5] Choma M A, Sarunic M V, Yang C, and Izatt J A 2003 *Opt. Exp.* **11** 2183–89

[6] Andrew M 1998 *Opt. Exp.* **3** 219–29
[7] Yaqoob Z, Wu J, and Yang C 2005 *Biotechniques* **39** S6–S13
[8] Povazay B, Bizheva K, Unterhuber A, Hermann B, Sattmann H, Fercher A F, Drexler W, Apolonski A, Wadsworth W J, Knight J C, Russell P St J, Vetterlein M, and Scherzer E 2002 *Opt. Lett.* **27** 1800–2
[9] Vakhtin A B, Peterson K A, and Kane D J 2006 *Opt. Lett.* **31** 1271–73
[10] De Boer J F, Cense B, Park B H, Pierce M C, Tearney G J, and Bouma B E 2003 *Opt. Lett.* **28** 2067–69
[11] Yun S H, Tearney G J, Bouma B E, Park B H, and de Boer J F 2003 *Opt. Exp.* **11** 3598–604
[12] Wojtkowski M, Kowalczyk A, Leitgeb R, and Fercher A F 2002 *Opt. Lett.* **27** 1415–18
[13] Drexler W 2004 *J. Biomed. Opt.* **9** 47–74
[14] Leitgeb R A, Drexler W, Unterhuber A, Hermann B, Bajraszewski T, Le T, Stingl A, and Fercher A F 2004 *Opt. Exp.* **12** 2156–65
[15] Fink M, Cassereau D, Derode A, Prada C, Roux P, Tanter M, Thomas J-L, and Wu F 2000 *Reports on Progress in Physics* **63** 1933-95
[16] Mittleman D M, Hunsche S, Boivin L, Nuss M C 1997 *Opt. Lett.* **22** 904–6
[17] Chen Y, Huang S, and Pickwell-MacPherson E 2010 *Opt. Exp.* **18** 1177–90
[18] Zaytsev K I, Karasik V E, Fokina I N, Alekhnovich V I 2013 *Opt. Eng.* **52** 068203
[19] Zaytsev K I, Chernomyrdin N V, Alekhnovich V I 2014 *J. Phys. Conf. Ser.* **486** 012010
[20] Fitzgerald A J, Berry E, Zinovev N N, Walker G C, Smith M A, and Chamberlain J M 2002 *Physics in Medicine and Biology* **47** R67–R84
[21] Podoleanu A G, Rogers j A, Jackson D A, and Dunne S 2000 *Opt. Exp.* **7** 292–8
[22] Wojtkowski M, Leitgeb R, Kowalczyk A, Bajraszewski T, Fercher A F 2002 *J. Biomed. Opt.* **7** 457–63
[23] Leitgeb R A, Schmetterer L, Drexler W, Fercher A F, Zawadzki R J, and Bajraszewski T 2003 *Opt. Exp.* **11** 3116–21
[24] Yasuno Y, Dimitrova Madjarova V, Makita S, Akiba M, Morosawa A, Chong C, Sakai T, Chan K-P, Itoh M, and Yatagai T 2005 *Opt. Exp.* **13** 10652–64
[25] Sergeev A M, Gelikonov V M, Gelikonov G V, Feldchtein F I, Kuranov R V, Gladkova N D, Shakhova N M, Snopova L B, Shakhov A V, Kuznetzova I A, Denisenko A N, Pochinko V V, Chumakov Y P, and Streltzova O S 1997 *Opt. Exp.* **1** 432–40
[26] Rollins A M, Yazdanfar S, Barton J K, Izatt J A 2002 *J. Biomed. Opt.* **7** 123–29
[27] Yang V X D, Gordon M L, Tang S J, Marcon N E, Gardiner G, Qi B, Bisland S, Seng-Yue E, Lo S, Pekar J, Wilson B C, and Alex Vitkin I 2003 *Opt. Exp.* **11** 2416–24
[28] Tumlinson A R, Hariri L P, Utzinger U, and Barton J K 2004 *Appl. Opt.* **43** 113–21
[29] Colston B W, Sathyam U S, DaSilva L B, Everett M J, Stroeve P, and Otis L L 1998 *Opt. Exp.* **3** 230–38
[30] Feldchtein F I, Gelikonov G V, Gelikonov V M, Iksanov R R, Kuranov R V, Sergeev A M, Gladkova N D, Ourutina M N, Warren J A, and Reitze D H 1998 *Opt. Exp.* **3** 239–50
[31] Fried D, Xie J, Shafi S, Featherstone J D B, Breunig T M, Le C 2002 *J. Biomed. Opt.* **7** 618–27
[32] Wilder-Smith P, Krasieva T, Jung W-G, Zhang J, Chen Z, Osann K, Tromberg B 2005 *J. Biomed. Opt.* **10** 051601
[33] Welzel J 2001 *Skin Research and Technology* **7** 1–9
[34] Kuranov R, Sapozhnikova V, Prough D, Cicenaite I, and Esenaliev R 2007 *Appl. Opt.* **46** 1782–86
[35] Drexler W, Fuijmoto J G *Optical Coherence Tomography: Technology and Applications* (Springer Publishing, 2008) 1400pp
[36] Alex A, Povazay B, Hofer B, Popov S, Glittenberg C, Binder S, and Drexler W 2010 *J. Biomed. Opt.* **15** 026025
[37] Mallat S G 1989 *IEEE Transact. Pattern Analysis and Machine Intelligence* **11** 674–93
[38] Antoine J-P, Carrette P, Murenzi R, and Piette B 1993 *Signal Processing* **31** 241–72
[39] Daubechies I 1990 *IEEE Transact. Information Theory* **36** 961–1005
[40] Torres J, Infante S O 2001 *Opt. Eng.* **40** 1309–14
[41] Federico A, Kaufmann G H 2001 *Opt. Eng.* **40** 2598–604
[42] Pie J, Ye P, and Xie W 2009 *Proc. SPIE* **7277** 727708
[43] Zaytsev K I, Chernomyrdin N V, Gorevoy A V, Trofimov N E, Fokina I N, Alekhnovich V I, Karasik V E, Yurchenko S O 2014 *J. Phys. Conf. Ser.* **486** 012034
[44] Chernomyrdin N V, Zaytsev K I, Gavdush A A, Fokina I N, Karasik V E, Reshetov I V, Kudrin K G, Nosov P A, Yurchenko S O 2014 *Proc. SPIE* **9216** 921611
[45] Wiener N *Time Series* (MIT Press, Cambridge, Massachusetts, USA, 1964) 42pp

Quality diagnostics of nanoscale AlAs/GaAs resonant tunnelling heterostructures based on IR-spectroscopic ellipsometry

Mstislav O Makeev*, Yury A Ivanov and Sergey A Meshkov
Bauman Moscow State Technical University, 105005, 2nd Baumanskaya str. 5, Moscow, Russia

E-mail: *mc.stiv@gmail.com

Abstract. In this work, the procedure of assessing the quality of AlAs/GaAs resonant tunnelling heterostructures with relation to their resistance to diffusion destruction was developed. The diffusion blur of the AlAs/GaAs heterostructure layers was detected by means of infrared-spectroscopic ellipsometry. The diffusion coefficients of Al and Si in GaAs were also calculated.

1. Introduction

The use of semiconductor devices based on the quantum size effects of charge transport is one possible means of improving the quality of radio electronic systems. Such devices are resonant tunnelling diodes (RTD) based on multilayer nanoscale AlAs/GaAs resonant tunnelling heterostructures with transverse charge transport. The short tunnelling time of the electrons provides extremely high performance of the diode (frequency up to 1 THz) and suppression of the shot noise [1].

By varying the parameters of the heterostructure layers (thickness, elemental composition), one can vary the I-V curve shape. Such properties of RTD allow one to create different nonlinear radio signal converters, including mixers, rectifiers, multipliers, and SHF and EHF generators [2-3]. There is an extensive bibliography devoted to studying RTD properties and issues with their technical radio applications, but the reliability of such devices requires further investigation. The problem of a reliability study of RTD is being solved by kinetics analysis of physical degradation processes in the RTD structure.

An RTD consists of a resonant tunnelling structure (RTS) (set of layers of AlAs/GaAs), contact regions (layers of Si-doped GaAs), and ohmic contacts. RTD I-V curve is highly sensitive to the changes of RTS parameters because RTS layers are very thin: only up to a few nanometers. Degradation processes (interlayer diffusion of Al in RTS and Si in contact regions) lead to changes in the shape of the potential barrier and potential well and, therefore, to changes of charge transport conditions in the structure (figure 1). Diffusion blur of ohmic contacts leads to an increase in their contact resistance. In turn, these processes lead to 'rifting' of the RTD I-V curve and deterioration of the output electrical characteristics of radio signal converters based on RTD.

The aim of the present work is the development of a technique allowing one to identify the diffusion coefficients of Al and Si in GaAs and to carry out quality diagnostics of resonant tunnelling heterostructures (RTHS) in relation to their resistance to diffusion destruction. Knowledge of the

Content from this work may be used under the terms of the Creative Commons Attribution 3.0 licence. Any further distribution of this work must maintain attribution to the author(s) and the title of the work, journal citation and DOI.
Published under licence by IOP Publishing Ltd

kinetics of diffusion processes in an RTD structure will allow one to predict reliability against gradual failures of RTDs and nonlinear radio signal converters.

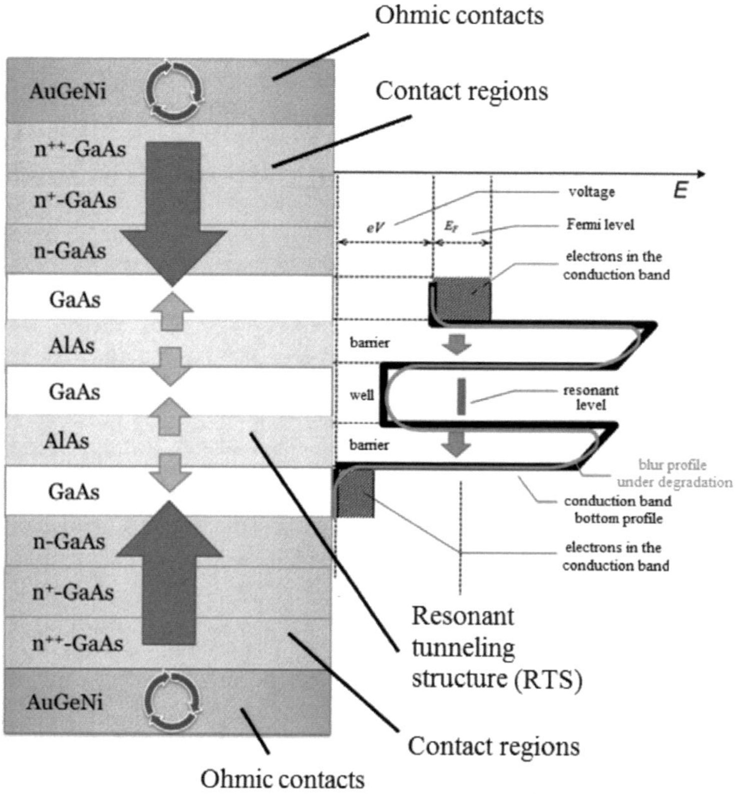

Figure 1. Structure and possible degradation phenomena. Yellow arrows: aluminium diffusion, blue arrows: silicon diffusion, circular red arrows: diffusion in ohmic contacts.

2. Results

The degradation of AlAs/GaAs RTHS as a result of Al diffusion in RTS and Si diffusion in the contact regions is considered. The intensity of diffusion processes in AlAs/GaAs RTHS is characterised by the diffusion coefficients of Al and Si in GaAs. Accordingly, quality assessment of RTHS (with relation to the resistance to diffusion destruction) is carried out based on the values of the diffusion coefficients of Al and Si in GaAs.

According to references [4-12], the diffusion coefficients of Al and Si in GaAs vary greatly depending on the conditions under which the heterostructures are produced by the means of molecular beam epitaxy (MBE). It is connected by the dependence of defectiveness of the growing heterostructures with respect to numerous technological factors such as substrate temperature (600–650 °C) during MBE, pressure in the chamber, temperature and time of annealing, defectiveness of the initial substrate, and many more.

The main diffusion mechanism for both Al and Si in GaAs is diffusion via vacancies of gallium (V_{Ga}). This is because atoms of Al and Si have similar masses and sizes and can be arranged in a Ga sublattice (with doping levels less than $5 \cdot 10^{18}$ cm^{-3}). Herein, diffusion occurs predominantly via negatively ionized vacancies of gallium with charge state ‘-3’. The diffusion coefficient depends also on the concentration of n-dopant according to the effect of Fermi level [4-8, 13-14].

The technique of nanoscale AlAs/GaAs RTHS diagnosis was developed for evaluation of diffusion coefficients. Structurally, it consists of the following main stages (figure 2): measurement of ellipsometric parameters of nanoscale AlAs/GaAs RTHS by the means of infrared-spectroscopic ellipsometry (IR-SE) before and after thermal influence, ellipsometric model creation, determination of profile distribution of Si in resonant tunnelling heterostructures, and calculation of diffusion coefficients on the basis of the determined profile.

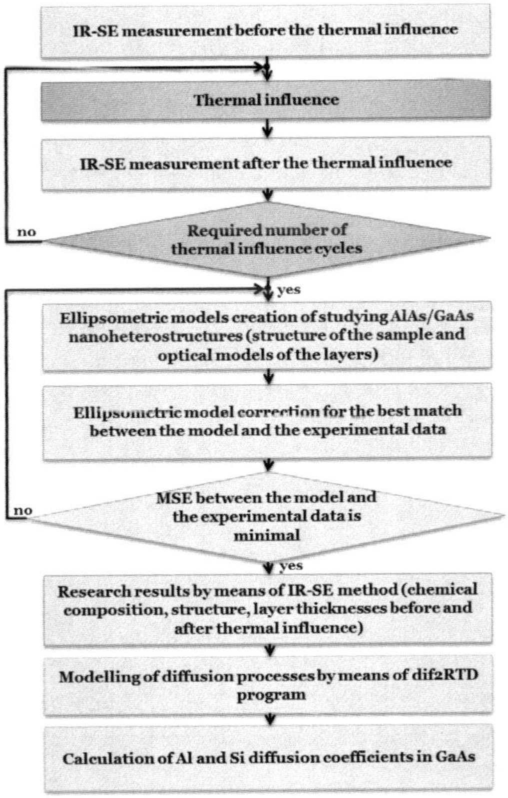

Figure 2. Block diagram of the quality diagnostics technique for nanoscale AlAs/GaAs resonant tunnelling heterostructures based on IR-spectroscopic ellipsometry.

The IR-spectroscopic ellipsometry technique is widely used for determination of thin film thicknesses and optical properties [15-18], semiconductor heterostructures [19-20], changes in chemical composition as the result of modification [21-22], etc.

Ellipsometric studies consist of the following stages: obtaining experimental data (spectra of Ψ and Δ), ellipsometric model creation (optical constants and layer thicknesses), and correction of model parameters for the purpose of obtaining the best fitting of experimental and generated data on the basis of this model obtaining the required results from the studies.

An ellipsometric model of the AlAs/GaAs RTHS samples includes the structure of the sample, optical constants, and thicknesses of all layers. The model is specified during research. Based on the specified ellipsometric model, one can estimate the chemical composition, structure, and layer thicknesses of AlAs/GaAs RTHS before and after thermal influence.

The calculation of the diffusion coefficients of Al and Si is carried out by the means of the dif2RTD program [23] for modelling diffusion processes in the nanoscale AlAs/GaAs RTHS and the I-V curve of RTD.

Temperature, time, and the number of thermal influences (artificial aging) and the number of measurements on the IR-spectroscopic ellipsometer are determined on the basis of prior data of chemical composition from studying nanoscale A_3B_5 heterostructure and the diffusion coefficients of the RTD heterostructure elements.

Validation of the quality diagnostics technique with regard to nanoscale AlAs/GaAs resonant tunnelling heterostructures based on IR-SE was carried out on two nanoscale heterostructures distinguished by layer structure and substrate temperature during epitaxial growth:

- GaAs (20 nm) / n-GaAs ($n_{Si} = 2 \cdot 10^{17}$ cm^{-3}, 100 nm) / GaAs (300 nm) / GaAs (substrate 450 μm) – Heterostructure № 1 (MBE temperature 650 °C);
- n-GaAs ($n_{Si} = 5 \cdot 10^{18}$ cm^{-3}, 100 nm) / n-GaAs ($n_{Si} = 2 \cdot 10^{17}$ cm^{-3}, 30 nm) / GaAs (1,5 nm) / AlAs (1,7 nm) / GaAs (4,5 nm) / AlAs (3,0 nm) / GaAs (10,0 nm) / n-GaAs ($n_{Si} = 2 \cdot 10^{17}$ cm^{-3}, 30 nm) / n-GaAs ($n_{Si} = 5 \cdot 10^{18}$ cm^{-3}, 100 nm) / GaAs (substrate 450 μm) – Heterostructure № 2 (MBE temperature 600 °C).

Thermal influence on heterostructure № 1 was carried out at a temperature of 530 °C and pressure $p = 10^{-4}$ Pa for 2 hours. Heterostructure № 2 was heated at a temperature of 300 °C for 4, 6, 8, and 10 hours (total annealing time was 28 hours) in the air environment.

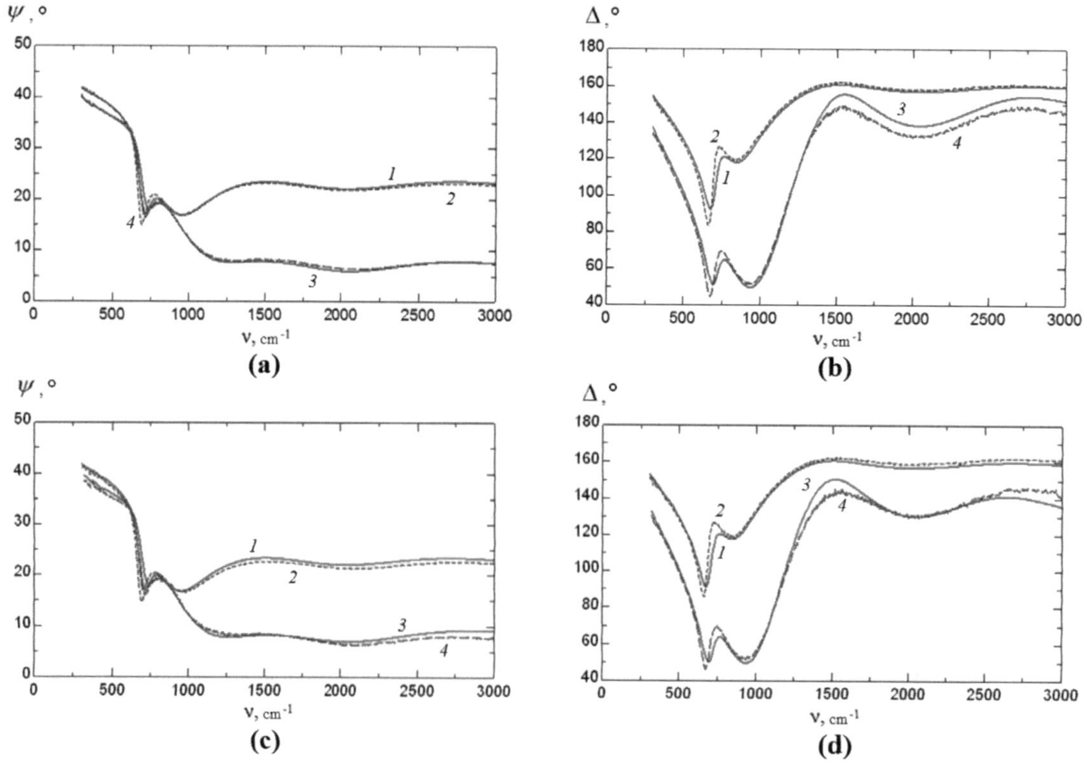

Figure 3. The spectra of the ellipsometric parameters Ψ (a, c) and Δ (b, d) of the AlAs/GaAs RTHS before (a, b) and after 28 hours (c, d) of thermal influence (*1* – generated data, the angle of incidence equals 60°; *2* – experimental data, the angle of incidence equals 60°; *3* – generated data, the angle of incidence equals 70°; *4* – experimental data, the angle of incidence equals 70°).

Ellipsometric model creation of the studied heterostructures, together with analysis of identified changes of the ellipsometric parameters Ψ and Δ before and after thermal influence (figure 3), allowed for the identification of diffusion blur of the layers belonging to the heterostructure (figure 4).

On this basis, diffusion coefficients of Al and Si in GaAs in heterostructure № 1 were calculated [20]:

$$D_{Al,Si} = 0{,}17 \cdot \exp\left(-\frac{3{,}5}{kT}\right) \cdot \left(\frac{n_{Si}}{n_i}\right)^3 \quad (1)$$

where $D_{Al,Si}$ = diffusion coefficients of Al and Si in GaAs,
k = Boltzmann constant ($k = 8{,}617 \cdot 10^{-5}$ eV·K^{-1}),
T = temperature, K,
n_{Si} = dopant carrier concentration (Si), cm^{-3},
n_i = intrinsic carrier concentration, cm^{-3}.

For heterostructure № 2 [21]:

$$D_{Al,Si} = 0{,}22 \cdot \exp\left(-\frac{3{,}5}{kT}\right) \cdot \left(\frac{n_{Si}}{n_i}\right)^3 \quad (2)$$

Top rough layer	24,9 ± 2,0 nm
n-GaAs heavily Si-doped (5 · 10^{18} cm^{-3})	100,3 ± 1,7 nm
Gradient layer	0 nm
n-GaAs Si-doped (2 · 10^{17} cm^{-3})	30,0 nm
Gallium arsenide GaAs (spacer)	1,5 nm
Aluminum arsenide AlAs (barrier)	1,7 nm
Gallium arsenide GaAs (well)	4,5 nm
Aluminum arsenide AlAs (barrier)	3,0 nm
Gallium arsenide GaAs (spacer)	10,0 nm
n-GaAs Si-doped (2 · 10^{17} cm^{-3})	30,0 nm
Gradient layer	0 nm
n-GaAs heavily Si-doped (5 · 10^{18} cm^{-3})	1005,0 ± 3,0 nm
Gradient layer	0 nm
Substrate GaAs	0,35 mm
Bottom rough layer	90,5 ± 10,0 nm

(a)

Top rough layer	15,9 ± 2,1 nm
n-GaAs heavily Si-doped (5 · 10^{18} cm^{-3})	103,4 ± 1,6 nm
Gradient layer	6,7 ± 1,5 nm
n-GaAs Si-doped (2 · 10^{17} cm^{-3})	30,0 nm
Gallium arsenide GaAs (spacer)	1,5 nm
Aluminum arsenide AlAs (barrier)	1,7 nm
Gallium arsenide GaAs (well)	4,5 nm
Aluminum arsenide AlAs (barrier)	3,0 nm
Gallium arsenide GaAs (spacer)	10,0 nm
n-GaAs Si-doped (2 · 10^{17} cm^{-3})	30,0 nm
Gradient layer	6,9 ± 1,8 nm
n-GaAs heavily Si-doped (5 · 10^{18} cm^{-3})	991,2 ± 3,6 nm
Gradient layer	7,1 ± 1,9 nm
Substrate GaAs	0,35 mm
Bottom rough layer	74,9 ± 10,0 nm

(b)

Figure 4. Ellipsometric models of the RTHS before (a) and after 28 hours (b) of thermal influence.

The pre-exponential factors in (1) and (2) differ by 30 %. Since the diffusion coefficients of Al and Si in GaAs are proportional to the amount of gallium vacancies [13-14, 24], one can conclude that the amount of defect in the studied heterostructures also differs by 30 %. It indicates that technological parameters (in this case MBE temperature) influence production quality (in the sense of the diffusion destruction processes rate) of RTHS. Thus, the IR-SE technique detects differences in the production quality of RTHS at different growing temperatures, and one can apply it for quality control.

3. Conclusions

The production quality of the AlAs/GaAs heterostructure of RTD was assessed in terms of the diffusion coefficient values of Al and Si in GaAs, representing the intensity of diffusion processes in the resonant tunnelling heterostructure.

In the present work, the possibility of using the IR-spectroscopic ellipsometry technique for calculation of the diffusion coefficients of Al and Si in AlAs/GaAs resonant tunnelling heterostructures was shown. Numerical characteristics of degradation processes, obtained in this work, can be used to predict the reliability of RTD and technical radio devices.

This technique developed on the basis of IR-spectroscopic ellipsometry allows one to 'see' the diffusion of the resonant tunnelling heterostructure AlAs/GaAs elements in nanometer layers. It allows one to determine diffusion coefficients of heterostructure elements. Since a diffusion coefficient directly depends on the defect amount in a heterostructure, one can assess the quality of more complex heterostructures and the value of their diffusion destruction in production process and in operation.

Acknowledgments
This work was carried out with the financial support from the Ministry of Education and Science of the Russian Federation (Mission № 16.1116.2014/K).

References

[1] Alkeev N V, Averin S V, Dorofeev A A, Gladysheva N B and Torgashin M Yu 2010 *Russ. Microelectronics* **39** 331
[2] Ivanov Y A, Meshkov S A, Fedorenko I A, Fedorkova N V and Shashurin V D 2010 *J. Commun. Technol. El.* **55** 982
[3] Ivanov Y A, Meshkov S A, Sinjakin V Y, Fedorenko I A, Fedorkova N V, Fedorov I B and Shashurin V D 2011 *Nanoengineering* **1** 34
[4] Yu S, Gösele U M and Tan T Y 1989 *J. Appl. Phys.* **66** 2952
[5] Chen C-H, Gösele U M and Tan T Y 1999 *Appl. Phys. A* **69** 313
[6] You H-M, Gösele U M and Tan T Y 1993 *J. Appl. Phys.* **73** 7207
[7] Pavesi L, Nguyen Hong Ky, Ganiere J D, Reinhart F K, Baba-Ali N, Harrison I, Tuck B and Henini M 1992 *J. Appl. Phys.* **71** 2225
[8] Bracht H, Haller E E, Eberl K and Cardona M 1999 *Appl. Phys. Lett.* **74** 49
[9] Wee S F, Chai M K, Homewood K P and Gillin W P 1997 *J. Appl. Phys.* **82** 4842
[10] Olmsted B L and Houde-Walter S N 1993 *Appl. Phys. Lett.* **63** 530
[11] Wang L, Hsu L, Haller E E, Erickson J W, Fischer A, Eberl K and Cardona M 1996 *Phys. Rev. Lett.* **76** 2342
[12] Ono H, Ikarashi N and Baba T 1995 *Appl. Phys. Lett.* **66** 601
[13] Fisher D J 1998 *Diffusion in GaAs and other III-V Semiconductors 10 Years of Research* (Zuerich: Scitec Publications)
[14] El-Mellouhi F and Mousseau N 2006 *J. Appl. Phys.* **100** 083521
[15] Hilfiker J, Bungay C, Synowicki R, Tiwald T, Herzinger C, Johs B, Pribil G and Woollam J A 2003 *J. Vac. Sci. Technol. A.* **21** 1103
[16] Hilfiker J, Singh N, Tiwald T, Convey D, Smith S M, Baker J H and Tompkins H G 2008 *Thin Solid Films* **516** 7979
[17] Tompkins H G, Tiwald T, Bungay C and Hooper A E 2004 *J. Phys. Chem. B.* **108** 3777
[18] Makeev M O and Zhukova E A 2013 *Science and education: electronic scientific and technical journal* **07** 229
[19] Makeev M O, Ivanov Y A and Meshkov S A 2011 *Nanoengineering* **4** 44
[20] Makeev M O, Ivanov Y A, Meshkov S A and Shashurin V D 2013 *Science and education: electronic scientific and technical journal* **11** 229
[21] Makeev M O, Ivanov Y A, Meshkov S A, Gil'man A B and Yablokov M Y 2011 *High Energ. Chem.* **45** 536
[22] Makeev M O, Ivanov Y A, Meshkov S A, Gil'man A B and Yablokov M Y 2011 *Nanotehnika* **3** 27
[23] Makeev M O, Ivanov Y A, Meshkov S A, Litvak Y N, Migal D E and Nazarov V V 2014 *Nanoengineering* **1** 24
[24] Mehrer H 2007 *Diffusion in solids: Fundamentals, methods, materials, diffusion-controlled processes* (Berlin: Springer)

Visible supercontinuum generation in large-core photonic crystal fiber with high air-filling fraction

Stanislav O Leonov*[1], Vladimir A Lasarev[1], Mikhail K Tarabrin[1], Dmitriy A Dvoretskiy[1], Alexandra S Pasishnik[2], Andrey D Pryamikov[3]
[1] Bauman Moscow State Technical University, 105005, 5/1, 2nd Baumanskaya str., Moscow, Russia
[2] S.I. Vavilov State Optical Institute, 192171, Babushkin str., 36/1, St. Petersburg, Russia
[3] Fiber Optics Research Center, Russian Academy of Sciences, 38, Vavilov str., Moscow, Russia

E-mail: *leonov-st@ya.ru

Abstract. We demonstrate new experimental results for visible supercontinuum generation in large-core photonic crystal fiber with high air-filling fraction. Supercontinuum is generated by picosecond pulses from a passively mode-locked Yb:KGW laser. A smooth supercontinuum generation near 800 nm is achieved using 4.5 m long large-core photonic crystal fiber even with 6 mW average coupled power. These results are of interest in high power white light sources and have potential applications in precise frequency metrology and spectroscopy. The data obtained could be used also in theoretical studies of nonlinear interaction in large-core photonics crystal fibers with high air-filling fraction.

1. Introduction

Photonic crystal fibers (PCFs) [1, 2] have drawn great scientific and technological interest since it was firstly demonstrated experimentally in 1996 [1]. PCFs have been employed for a wide range of purposes, among which are supercontinuum (SC) sources [3, 4] based on femtosecond lasers [5, 6], spectroscopic gas analysis of atmosphere [7], optical coherence tomography [8], optical frequency metrology [9-11] etc. A lot of research was devoted to SC generation in a short lengths of endlessly single-mode small-core PCFs pumped by femtosecond laser pulses [10] or several meter-long PCFs pumped by picosecond sources [13-15]. Using large mode area (LMA) PCFs in comparison with small-core ones it is possible to achieve low numerical aperture [16] and only fundamental mode will propagate at all wavelengths [17]. SC generation in single-mode LMA PCFs using nanosecond pulse pumping with long-wavelength emission edge at 1.7 μm was demonstrated in [18]. The visible part of SC was rather low and was observed only at maximum coupled power. Also the SC generation in a multimode LMA PCFs is of great interest. The infrared (IR) SC generation in a multimode LMA PCFs with high air-filling fraction under extreme (close to fiber damage threshold) picosecond excitation with long-wavelength emission edge at 1.9 μm was discussed in [19, 20]. At the same time the visible (VIS) SC generation in the LMA PCFs with high air-filling fraction under picosecond excitation is still of interest to obtain [21].

Content from this work may be used under the terms of the Creative Commons Attribution 3.0 licence. Any further distribution of this work must maintain attribution to the author(s) and the title of the work, journal citation and DOI.
Published under licence by IOP Publishing Ltd

In this article, we demonstrate the new experimental results for the VIS SC generation in large-core multimode photonic crystal fiber with high air-filling fraction pumped with ps pulses from Yb:KGW laser.

2. Experiments

The fiber LC-PCF-10 used in our experiments is 4.5 m long large-core multimode photonics crystal fiber with a triangular microstructured cladding [22] and four cycles of air holes. The fiber core diameter is equal to 10 μm. An optical microscope picture of the cross section is shown in figure 1 (a).

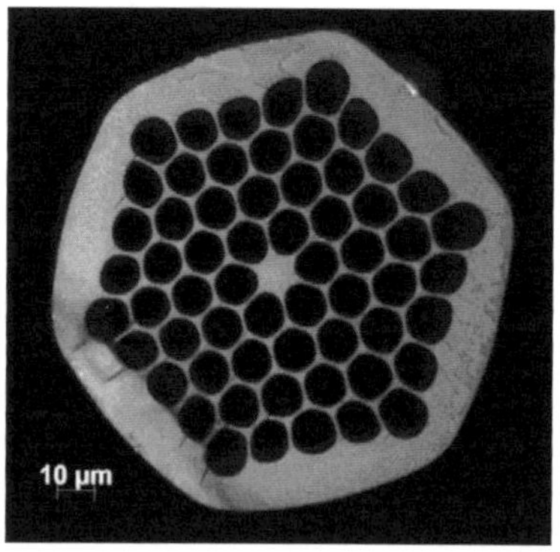

a)

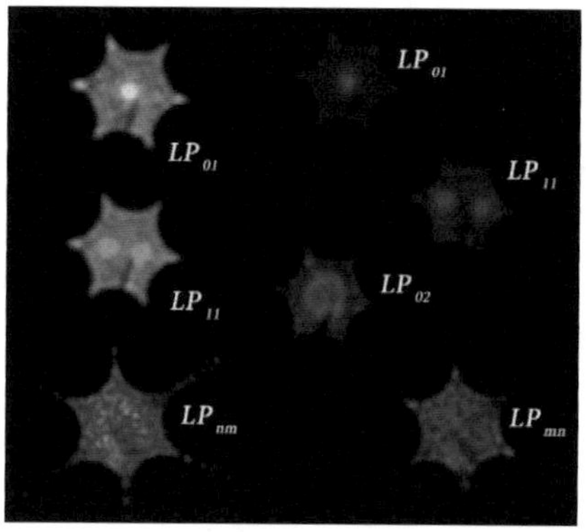

b)

Figure 1. Optical microscope picture of the cross section of the multimode large core PCF (a). Visualized modes at the output fiber edge (b).

The air-filling fraction of the fiber that is the ratio of the capillary hole diameter d to the pitch of structure Λ is approximately equal to 0.9, which means that the fiber supports multiple modes from visible to infrared wavelength ranges [23]. The multimode guidance was verified both theoretically using COMSOL and experimentally by launching light at different wavelengths (550 and 650 nm) into the fiber as it is shown in figure 1 (b). The characteristics of the LC-PCF-10 are summarized in Table 1.

Table 1. Characteristics of the LC-PCF-10 for fundamental mode

	Core Diameter (μm)	Λ (μm)	d/Λ	Fiber Length (m)	ZDW (nm)	γ (W^{1}/km)	A_{eff} (μm^2) @Pump
LC-PCF-10	10	11	0.9	4.5	1175	2.324	52.5

The dispersion profile of fiber was also calculated using COMSOL and is plotted in figure 2 (b). It is easy to notice that zero dispersion wavelength (ZDW) for LP_{01} and LP_{02} modes are different. For LP_{02} ZDW is shifted to 1 μm. The calculated value of effective mode area is rather large, so the nonlinear coefficient of the fiber is lower in comparison with small-core PCFs. Nevertheless, the lower nonlinear coefficient can be compensated by an increase of input power of laser radiation or fiber length.

An experimental setup is illustrated in figure 2 (a). A laser source is employed by Yb:KGW laser (Pharos-4W) with central wavelength 1026 nm. The laser operates at repetition rate of 200kHz and produces pulses with temporal width of ~250 fs. The maximum average output power of the laser was 4 W. The initial femtosecond pulses were stretched with diffraction gratings pair so the temporal length of output pulses was 2 ps. Experimentally, pulses from Yb:KGW laser is coupled into a 4.5 m long piece of the photonic crystal fiber with 50x infinity corrected lens with focal length 4 mm (Mitutoyo). The laser radiation was focused in a spot about 2 μm on the fiber edge. The output SC radiation from fiber was simultaneously measured by two optical spectrum analyzers VIS (Thorlabs) and IR (Avantes). A special Si cutting filter was placed before the IR spectrometer input in order to eliminate influence of the second order diffraction of SC generation visible part.

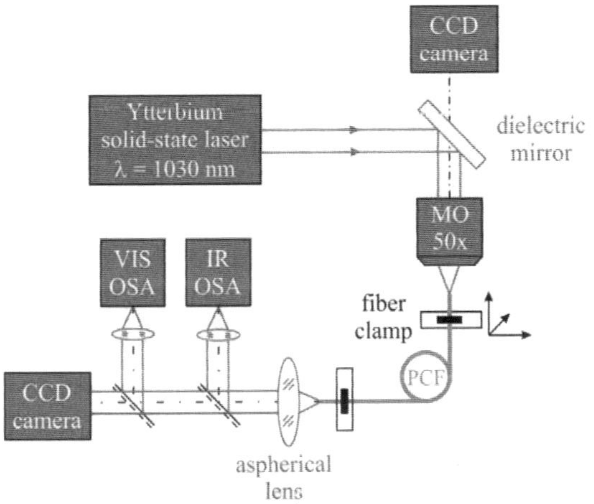

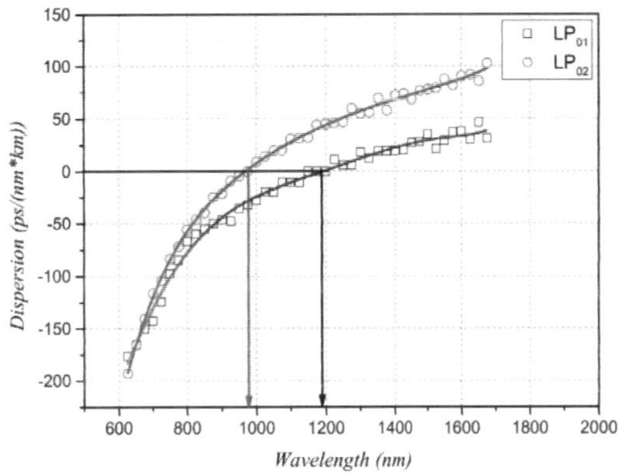

Figure 2. Experimental setup (a). LC-PCF-10 dispersion profile (b).

The SC spectrum generated in LC-PCF-10 is shown in figure 3 for increasing average coupled power. The exited fiber mode is presented on inset in figure 3.

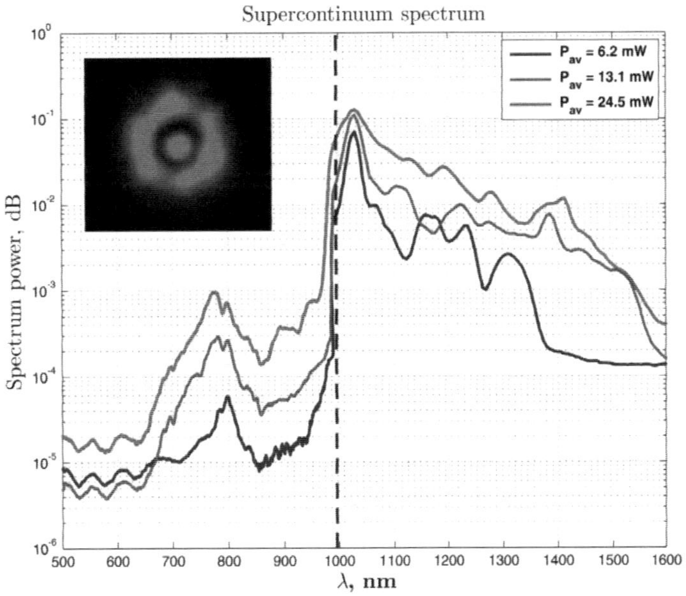

Figure 3. Supercontinuum generation in LC-PCF-10 fiber. P_{av}: average power measured at the fiber output. Inset: Output fiber mode. The dashed line marks the location of zero-dispersion wavelength for LP_{02} fiber mode.

Figure 3 shows that visible spectrum components appear even with 6 mW coupled power. Increasing coupled power leads to the spectral broadening mostly into infrared region. At maximum input power the spectrum components at 900 nm start to generate and infrared part of the spectrum becomes smoother. Such dependence allows managing the SC spectra by increasing the input power to 100 mW or more.

3. Conclusions

In summary, we have demonstrated, for what is believed to be the first time, visible supercontinuum generation in large-core photonic crystal fiber with core diameter of 10 μm and air-filling fraction of 0.9 under picosecond excitation by 2 ps pulses at 1.03 μm. The visible SC generation was obtained for LP_{02} mode due to pump central wavelength was closed to zero dispersion wavelength of LP_{02}. The visible spectrum component was observed even with 6 mW coupled power. These results are of immediate interest in high power white light sources and have potential applications in precise frequency metrology and spectroscopy.

References

[1] Knight J C, Birks T A, Russell P St J and Atkin D M 1996 *Opt. Lett.* **21** 1547-49
[2] Russel P St J 2003 *Science* **299** 358-62
[3] Dudley J M, Genty G and Coen S 2006 *Rev. Mod. Phys.* **78** 1135-84
[4] Fedotov A B, Naumov A N, Zheltikov A M, Sugar I, Chorvat J D, Tarasevitch A P, Von der Linde D 2002 *J. Opt. Soc. Am. B* **19** 2156-64
[5] Fermann M E, Hartl I 2013 *Nature Photonics* **7** 868–874
[6] Liu Y, Zhao X, Liu J, Hu G, Gong Z, Zheng Z 2014 *Opt. Exp.* **22** 21012-17
[7] Swartling J, Bassi A, D'Andrea C, Pifferi A, Torricelli A, and Cubeddu R 2005 *Appl. Opt.* **44** 4684-92
[8] Hartl I, Li X D, Chudoba C, Ghanta R K, Ko T H, Fujimoto J G, Ranka J K, and Windeler R S 2001 *Opt. Lett.* **26** 608-10
[9] Kadulova M, Hlubina P, Buczynski R 2013 *Proc. of SPIE* **8775** 87750X
[10] Holzwarth R, Udem Th, Hansch T W, Knight J C, Wadsworth W J and Russell P. St. J. 2000 *Phys. Rev. Lett.* **85** 2264-67
[11] Jones D J, Diddams S A, Ranka J K, Stentz A, Windeler R S, Hall J L and Cundiff S T 2000 *Science* **288** 635-39
[12] Ranka J K, Windeler R S and Stentz A J 2000 *Opt. Lett.* **25** 25-27
[13] Coen S, Hing Lun Chau A, Leonhardt R, Harvey J D, Knight J C, Wadsworth W J, and Russell P St J 2001 *Opt. Lett.* **26** 1356-58
[14] Wadsworth W J, Joly N, Knight J C, Birks T A, Biancalana F and Russell P St J 2004 *Opt. Exp.* **12** 299-309
[15] Dudley J M, Provino L, Grossard N, Maillotte H, Windeler R S, Eggleton B J and Coen S 2002 *J. Opt. Soc. Am. B* **19** 765-71
[16] Issa N A 2004 *Appl. Opt.* **43** 6191–97
[17] Knight J C, Birks T A, Cregan R F, Russell P St J and de Sandro J P 1998 *Electron. Lett.* **34** 1347-48
[18] Genty G, Ritari T and Ludvigsen H 2005 *Opt. Exp.* **13** 8625
[19] Savitski V G, Yumashev K V, Kalashnikov V L, Shevandin V S, Dukel'skii K V 2007 *Opt. Quant. Electron.* **39** 1297
[20] Savitski V G, Yumashev K V, Kalashnikov V L, Shevandin V S, Dukel'skii K V 2007 *Proc. of SPIE* **6729** 672914-1
[21] Pasishnik A S, Leonov S O 2014 *Proc. of SPIE* **9136** 91361F
[22] Demidov V V , Dukel'skii K V, Pasishnik A S , Shevandin V S 2013 *WSOF Conf. Paper* **F2** F2.18
[23] Nielsen M D and Mortensen N A 2003 *Opt. Exp.* **11** 2762-68

Journal of Physics: Conference Series **584** (2015) 012016 doi:10.1088/1742-6596/584/1/012016

Mathematical analysis of marine pipeline leakage monitoring system based on coherent OTDR with improved sensor length and sampling frequency

A B Pnev, A A Zhirnov, K V Stepanov, E T Nesterov, D A Shelestov and V E Karasik
Bauman Moscow State Technical University, 2-nd Baumanskaya st. 5, 105005 Moscow, Russia

E-mail: zhirnovrl@gmail.com

Abstract. A system based on coherent optical time domain reflectometry (OTDR) for subsea pipeline monitoring is described. The fiber sensor length is increased using erbium-doped fiber amplifier (EDFA) cascades. The sampling frequency is increased by dividing the fiber sensor into separate sensitive areas, with parallel scanning. The calculation of the erbium amplifier cascade spontaneous noise influence on the signal-to-noise ratio (SNR) is carried out.

1. Introduction

Hydrocarbons, in particular gas and oil, are some of the most in demand goods at present. The main transport systems for hydrocarbon delivery are pipelines. Their length can vary from a few hundred meters to several thousand kilometers. They can be installed both on land and at the bottom of the sea. More than one hundred leaks are found in pipelines in Russia every year. According to experts, that number is going to increase significantly in the future. According to information from Greenpeace, Russia annually loses 10 to 20 million tonnes of oil and 6 to 50 billion m3 of gas due to leaks. Such a leakage volume represents from 3% to 7% of the total transported oil or gas. All of these hydrocarbons eventually pollute the environment. According to calculations, up to 24 billion m^3 of gas can be saved by reducing the number of leaks to international standards.

The occurrence of leaks also leads to environmental pollution by hydrocarbons, which have a strong negative impact on the biosphere. The importance of leak detection is obvious. The economically efficient use of pipelines requires early detection by means of a monitoring system.

This paper is organized as follows. In Section 2, we briefly describe the most common methods of leak detection and introduce a fiber optic system for pipeline monitoring. In Section 3, we show disadvantages of this system. In Section 4, we propose a method to eliminate them and theoretically calculate SNR of the considered system. Section 5 concludes the main results of the paper.

2. Methods

The monitoring system must satisfy a number of requirements: high sensitivity, high accuracy for determining the location of the leak, safe and reliable long-term use, possibility for long-distance pipeline control, high degree of automation, absence of interruption in the pumping process, efficiency, and a capability to work in any climate and weather conditions.

Content from this work may be used under the terms of the Creative Commons Attribution 3.0 licence. Any further distribution of this work must maintain attribution to the author(s) and the title of the work, journal citation and DOI.
Published under licence by IOP Publishing Ltd

Existing methods can be classified into two large groups: dynamic and static. The first group does not require stopping product transportation, and the second group does. Existing dynamic control methods have the capability to detect leakages of more than 50 L/h. Smaller leaks can only be detected by static methods. Each method has its own set of parameters: sensitivity, scanning speed, positioning accuracy, etc. The most widespread techniques are as follows: wave alert, mass balance, and acoustic emission systems [1]. The first and second methods cannot detect small leaks. The last method can, but, at present, it requires a special staff with acoustic detectors that can scan only a few kilometers of pipeline per day. These disadvantages impair the automation of the process, prolong the inspection time, and reduce the rate of detection.

According to Riemsdijk et al., [2] the sound pressure level of a leak is about 120 dB. This level is comparable to that from the passage of heavy trucks; hence, it is possible to detect leaks not only by means of point sensors, but also by using distributed fiber-optic sensors. In this paper, we propose a system that allows continuous monitoring of acoustic signals along the entire length of the pipeline, in a fully automated mode. It is based on the principle of coherent optical time domain reflectometry (OTDR). In this method, the optical signal that propagates along the fiber is partially attenuated and scattered by inhomogeneities. Part of the radiation scattered within the fiber is sent backwards. By recording and analyzing the backscattered signal, we can recover losses distributed along the length of the fiber.

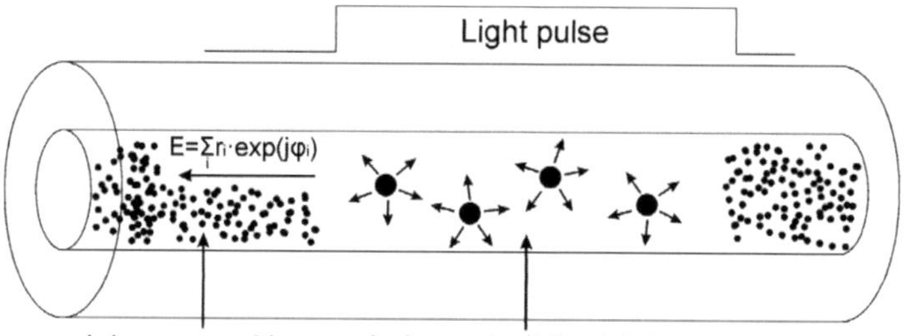

Figure 1. Rayleigh scattering in an optical fiber.

If the coherence length of the laser is greater than the pulse width, we can see the effect of coherent scattering of light, and the photoreceiver measures summed amplitudes of the backscattered waves. The total amplitude of that signal can be represented as a sum of vectors with random amplitude and phase. For coherent scattering of light, the total amplitude of the scattered wave varies randomly along the fiber from its maximum value to zero, due to interference. Consequently, relative power fluctuations of the scattered radiation are approximately equal to one, and a signal (reflectogram) detected by a photoreceiver undergoes significant deviations. If something impacts the fiber at a given place, the interference pattern changes at this location because of phase modulation of the backscattered wave from that region. It causes changes in the reflectogram. Systems, based on this method, attract a significant research attention [3-11].

3. Existing parameters

Application of such systems for industrial leak detection is only now in the development stage. One of the tasks necessary to improve the system quality is matching the acoustic bands of the leak emission (10-40 kHz) and the sampling frequency of the coherent OTDR:

$$f_{samp} = \frac{c}{2L_{sens}n} = 2\ kHz \tag{1}$$

where c - speed of light in vacuum, L_{sens} - fiber sensor length, and n - refractive index of the fiber. For typical values L_{sens} = 50 km and n = 1.5. This means that we have to increase the sampling frequency limit by at least a factor of 10. As can be seen from equation (1), these target values can be achieved only by reducing the length of the sensor. This would be a problem because the pipelines may have lengths of several hundred kilometers, so the length of the fiber sensor is a critical issue.

4. Proposed solution and theoretical calculation

As shown in figure 2, a system has been proposed to combine the advantages.

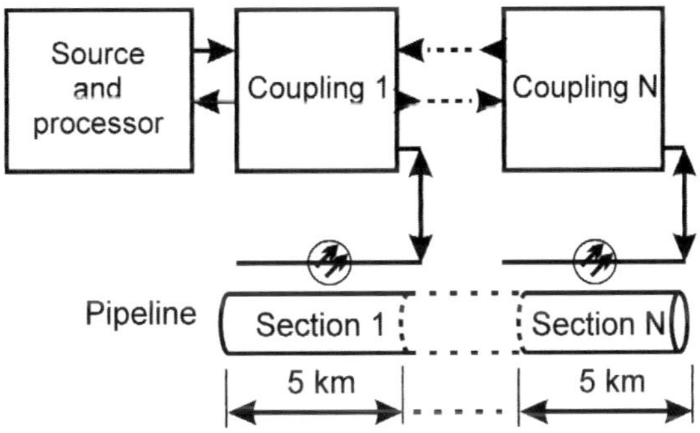

Figure 2. Scheme of proposed COTDR.

In this scheme, the pulse from the source node propagates to each branch, where a fraction of the radiation is sent to the sensor section and another part is amplified and transmitted to the next coupling. Thus, amplification of the pulse, propagating through the couplings, by the EDFA simultaneously reduces the sensor section length and increases the sampling frequency. The optimum length of the fiber section should be equal to the submarine cable installation length, which, at present, is 5 km. This will provide both ease of system assembly and the desired range for acoustic frequencies, because in this scheme $f_{samp} = 20\ kHz$.

The disadvantage of a system with a large number of branches and, consequently, erbium amplifiers, is spontaneous emission, which reduces the SNR. Calculations of couplers' influence on the SNR of the system are described in two stages:

1. definition of SNR_1 for one sensor section;
2. definition of the deterioration of the SNR due to the amplified spontaneous emission (ASE) from cascades of N erbium amplifiers.

In the first stage, we calculate the SNR of a system without EDFA, constructed as shown in figure 3.

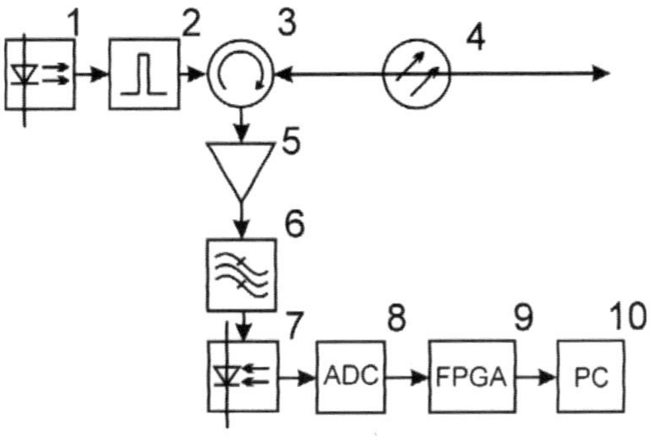

Figure 3. Scheme without EDFA. The components are: 1 - narrow bandwidth laser, 2 – acousto-optic modulator, 3 - optical circulators, 4 - fiber sensor, 5 - optical pre-amplifier, 6 - optical filter, 7 - photoreceiver, 8 - analog-digital converter, 9 - field-programmable gate array (FPGA), and 10 – personal computer.

The maximum input power for this scheme is limited by nonlinear effects (primarily by self-phase modulation) [12], so we choose the source power, P_{in} =200 mW = 23 dBm. The SNR of a system without amplifier cascades is analyzed as follows:

$$SNR_1 = 10lg\left(\frac{P_{out}}{\sqrt{\sigma_T{}^2 + \sigma_n{}^2 + \sigma_A{}^2 + \sigma_{S-ASE}{}^2 + \sigma_{ASE-ASE}{}^2}}\right) \approx 17.5\ dB \qquad (2)$$

where P_{out} - the power obtained on the photoreceiver, σ_T - the variance of receiver thermal noise, σ_n - the variance of receiver shot noise, σ_A - the variance of electrical amplifier noise, σ_{S-ASE} - the variance of signal-ASE beat noise in an optical pre-amplifier, and $\sigma_{ASE-ASE}$ - the variance of ASE-ASE beat noise in optical pre-amplifier.

In the second stage, the SNR is calculated for a system with EDFA cascades, as shown in figure 4. Effects from all the cascades are combined to one equivalent cascade (3, 4).

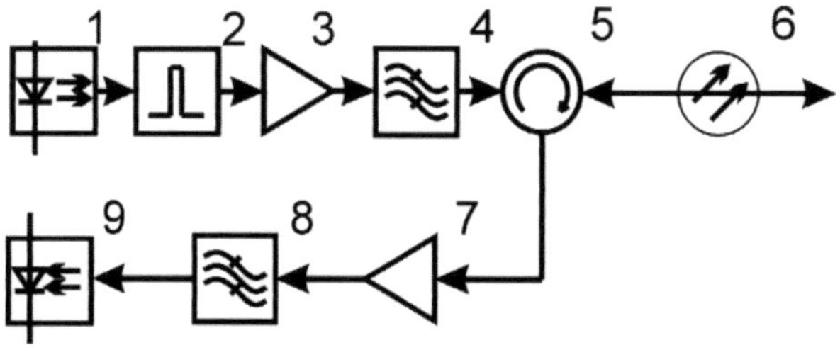

Figure 4. Equivalent cascade scheme. 1 - narrow bandwidth laser, 2 – acousto-optic modulator, 3 - EDFA booster, 4, 8 - optical filter, 5 - optical circulators, 6 - fiber sensor, 7 - optical pre-amplifier, 9 - photoreceiver.

We have to define the influence ASE of these EDFA on the SNR. A single-cascade is shown in figure 5. The attenuation coefficient for a 5-km-long sensor section equals $K_{att} = \alpha L$ = 0.17 dB/km×5 km = 0.85 dB. Furthermore, we use the coupler (K_{div} = 99% to sensor), so the reduction of ASE on the next cascade is achieved.

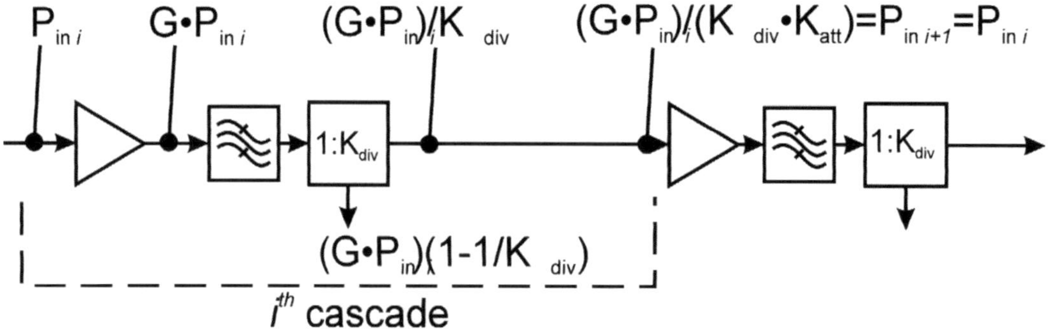

Figure 5. i[th] cascade scheme.

Based on the condition that the input power on every stage must be equal ($P_{in\ i} = P_{in\ i+1}$), we can estimate the EDFA gain $G = K_{div}K_{att} \approx 120$. Finally, we find the equation for the output pulse power of the Nth cascade, considering the ASE of previous EDFA by induction:

$$P_{out\ N} = K_{att}\left(\frac{G}{K_{div}K_{att}}\right)^N P_{in} + \frac{P_{ASE}}{K_{div}}\sum_{m=0}^{N-1}\left(\frac{G}{K_{div}K_{att}}\right)^m = K_{att}P_{in} + \frac{N}{K_{div}}P_{ASE} \qquad (3)$$

From the obtained expression, one can see that the power of the background noise will increase with the number of cascades and does not depend on the propagating pulse power. After computation, the power of ASE was estimated as:

$$P_{ASE} = (NF\ G - 1)h\nu\Delta\nu = 1.915 \times 10^{-6}\ W \quad (4)$$

where $NF = 4$ - pre-amplifier noise figure, $G = 100$ - pre-amplifier gain, h - Planck's constant, $\nu = 193.4$ THz - radiation frequency, and $\Delta\nu = 12.5$ GHz - optical pre-amplifier bandwidth, we obtained the noise power as a function of the number of amplifiers:

$$P_{noise} = \frac{N}{K_{div}} P_{ASE} G - N \times 2.3 \times 10^{-6}\ [W] \quad (5)$$

In equation (2), the pulse power fluctuations produce the major contribution to the signal-ASE beat noise:

$$\sigma_{s-ASE}{}^2 = 2S^2G^2(P_{sig} + P'_n)NFh\nu\Delta f \quad (6)$$

where $\Delta f = 5$ MHz - receiver bandwidth, and

$$\sigma_{s-ASE}{}^2 = 2S^2G^2(P_{sig} + P'_n)NFh\nu\Delta f \quad (7)$$

- Rayleigh scattering from the ASE of all EDFA in cascades, calculated as shown in [13], where L - losses in the line. The term $10\lg(1/f_{samp})$ takes into account the fact that the ASE is generated continuously, not only during the pulse width τ. Because the signal power, $P_{out} \approx 10^{-5}$ W, one can see that even for a total sensor length of 200 km (corresponding to $N = 40$ cascades), ASE does not significantly affect the SNR:

$$SNR = 10log(P_{out}/(\sigma_T{}^2 + \sigma_n{}^2 + \sigma_A{}^2 + (\sigma_{s-ASE}(1 + 0.0018N))^2 + \sigma_{ASE-ASE}{}^2)^{1/2}) \approx 17\ dB \quad (8)$$

Figure 6 shows a detailed schematic of the final technical solution proposed in this paper and we provide a description of the device's operation. Radiation from the source, 1, acquires pulse modulation in the acousto-optic modulator (AOM), 2. In the optical amplifiers, 3, it gains the level required for entry into the fiber sensor and transmission to the next section. The optical filter, 4, reduces the spontaneous emission. The signal then enters the optical coupler, 5, from which one part of the pulse is transmitted to the next sections, and another part enters the current branch. Before the sensor itself, a small part of the radiation is routed by the coupler, 6, to an auxiliary receiver, 7.1, for pulse start synchronization. The main part of the power entering the circulator, 8, is directed to the sensor, 9, and then the backscattered radiation is directed by the circulator to the pre-amplifier, 10, with an optical filter, 4, to increase signal power, and is detected by the receiver, 7.2. Next, the field-programmable gate array (FPGA), 12, controls the analog-digital converter (ADC), 11, that translates the signal into a digital form that passes through the communication devices, 13, to the host processor, located at the station on the coast.

We now focus on some of the practical features of this scheme. Devices 3, 4, 5, 6, 7.1, 7.2, 8, 9, 11, 12, and 13 are placed in small waterproof couplings. These couplings are standard telecommunication devices. Couplings are united in pairs with one common EDFA to decrease its influence on the pulse form. Electrical power is supplied by a special marine cable, which lies along an underwater pipeline.

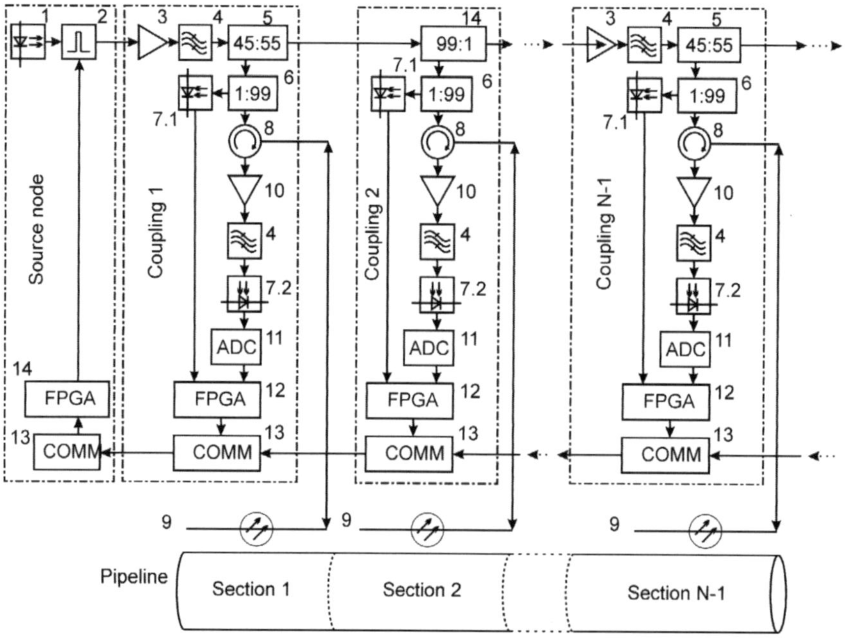

Figure 6. Scheme of proposed system. The components are: 1 - narrow bandwidth laser, 2 – acousto-optic modulator, 3 - EDFA booster, 4 - optical filter, 5, 6 - optical couplers, 7.1, 7.2 - photoreceivers, 8 - optical circulators, 9 - fiber sensor, 10 - optical pre-amplifier, 11 - analog-digital converter, 12, 14 - field-programmable gate array (FPGA), and 13 - communications devices.

5. Discussion and conclusions

A system based on a cascade coherent OTDR was developed that was capable of registering acoustic emission signals from leaks in pipelines. The range of detected sound frequencies was increased to 10 kHz using a cascade scheme to implement coherent OTDR. It was shown that the spontaneous emission noise of the amplifiers for a 200-km sensor array does not have a significant effect on the SNR of the system.

Acknowledgments
This work is supported by Russian Ministry of Industry and Trade, State Contract № 13411.1007499.09.033.

References

[1] Stafford M and Williams N 1996 *Pipeline leak detection study* (London: HSE Books)
[2] Riemsdijk A J and Bosselaar H 1970 *Erdoel – Erdgas – Zeitschrift* **86** 12-8
[3] Taylor H F and Lee C E 1993 *US 5194847 A*
[4] Qin Z, Zhu T, Chen L and Bao X 2011 *IEEE Photonics Technol. Lett.* 23 1091-3
[5] Bao X and Chen L 2012 *IEEE Sensors J.* 12 8601-8639
[6] Peng F 2014 *Optical Fiber Communication Conference* pp. M3J-4
[7] Peng F, Wu H, Jia X H, Rao Y J, Wang Z N and Peng Z P 2014 *Opt. Express* **22**, 13804-10
[8] Juarez J C and Taylor H F 2007 *Appl. Opt.* **46** 1968-71
[9] Zhu T, He Q, Xiao X and Bao X 2013 *Opt. Express* **21** 2953-63
[10] Parker T, Shatalin S and Farhadiroushan M 2014 *First Break* **32** 61-9
[11] Alekseev A E, Tezadov Y A and Potapov V T 2013 *Tech. Phys. Lett.* **39** 42-45
[12] Izumita H, Koyamada Y, Furukawa S I and Sankawa I 1994 *J. Lightwave Technol.* **12** 1230-8
[13] Listvin A V and Listvin V N 2005 *Reflectometry of optical fibers* (Moscow: LESARart)

Hyper-spectral modulation fluorescent imaging using double acousto-optical tunable filter based on TeO_2-crystals

Kirill I Zaytsev[1], Alexey V Perchik[1], Nikita V Chernomyrdin[1], Konstantin G Kudrin[2], Igor V Reshetov[2] and Stanislav O Yurchenko[1]

[1]Bauman Moscow State Technical University, 105005, 2nd Baumanskaya str. 5, Moscow, Russia;
[2]P.A.Hertzen Moscow Research Institute of Oncology, 125284, 2nd Botkinsky proezd 3, Moscow, Russia

E-mail: `kirzay@gmail.com`

Abstract. We have proposed a method for hyper-spectral fluorescent imaging based on acousto-optical filtering. The object of interest was pumped using ultraviolet radiation of mercury lamp equipped with monochromatic excitation filter with the window of transparency centered at 365 nm. Double TeO_2-based acousto-optical filter, tunable in range from 430 to 780 nm and having 2 nm bandwidth of spectral transparency, was used in order to detect quasi-monochromatic images of object fluorescence. Modulating of ultraviolet pump intensity was used in order to reduce an impact of non-fluorescent background on the sample fluorescent imaging. The technique for signal-to-noise ratio improvement, based on fluorescence intensity estimation via digital processing of modulated video sequence of fluorescent object, was introduced. We have implemented the proposed technique for the test sample studying and we have discussed its possible applications.

1. Introduction

Since first appearance of various fluorescent imaging techniques they have attracted considerable interest as possible methods of fundamental and applied research. They provide an ability for studying biological objects, single particles and cells, for visualizing structure and movements of object [1–8].

The methods of fluorescent imaging [8–16], fluorescent microscopy [17–22], and fluorescent tomography [23–25] have been developed during the last decades. Many approaches for fluorescent image processing, for tomographic reconstruction of the sample structure based on its fluorescent imaging were proposed, and many technical realizations of multi-spectral ultra-sensitive fluorescent image detection were developed. These techniques were successfully implemented for the purpose of medical diagnosis, including diagnosis of skin cancers [26,27]. However, the problem of development of hyper-spectral continuously-tunable fluorescent imaging systems, and the problem of fluorescent image processing remain very important for further progress in fluorescent imaging applications.

Content from this work may be used under the terms of the Creative Commons Attribution 3.0 licence. Any further distribution of this work must maintain attribution to the author(s) and the title of the work, journal citation and DOI.
Published under licence by IOP Publishing Ltd

This paper presents the novel results of modulation fluorescent imaging system development. We are describing the fluorescent imaging system based on pumping the sample with modulated ultraviolet (UV) radiation of mercury lamp, equipped with 365 nm excitation filter. Double TeO_2-based acousto-optical filter, tunable in range from 430 to 780 nm and having 2 nm bandwidth of spectral transparency, is used for detecting the quasi-monochromatic images of fluorescent object. We utilize the pumping intensity modulation simultaneously with the technique for digital processing of video sequence of modulated fluorescent images, in order to achieve high signal-to-noise (SNR) ratio, as well as to reduce an impact of non-fluorescence background on the results of fluorescent measurements. We are implementing the proposed technique for the test sample studying; the results of experiments and possible applications of proposed method are discussed.

2. Materials and methods

At first we are describing the experimental set-up for fluorescence imaging. Figure 1 shows the schematic representation of experimental set-up. We are utilizing the UV radiation of mercury lamp with 100 W power of polychromatic beam to pump the sample, and we use excitation filter transparent around $\lambda_p = 365 \pm 5$ nm. The intensity of UV beam is modulated using mechanical modulator (beam chopper); thus, the time-domain intensity profile has the following character

$$I(t) \sim 1 + sin\left(2\pi\nu_p t\right), \tag{1}$$

where $\nu_p = 2.0$ Hz is a frequency of modulation, and the modulation depth is $m = 1.0$. Being radiated with UV, the sample fluoresces, and the intensity of fluorescence is also modulated with the same frequency, ν_p. The video image of sample fluorescence is being detected by means of complementary metaloxidesemiconductor (CMOS) camera equipped with double acousto-optical tunable filter (DAOTF) [28,29]. The frame rate of CMOS camera is much higher than the UV beam modulation frequency, $\nu_{frame} = 25.0$ Hz $\gg \nu_p$. Thus, the video sequence of sample fluorescence represents blinking of fluorescence intensity, allowing us to estimate the spatial distribution of fluorescence intensity by means of processing the video sequence.

The main feature of presented set-up is associated with the presence of DAOTF [28,29]. It consists of two single acousto-optical filters (AOF) based on TeO_2 crystals, three $CaCO_3$ polarizers and additional parts for acoustic wave generation and control. If linearly polarized electromagnetic wave is propagating througth the TeO_2-crystal simultaneously with the acoustic wave, and if the condition of non-collinear light diffraction is satisfied

$$\mathbf{k}_s = \mathbf{k}_i + \mathbf{Q}, \tag{2}$$

where $\mathbf{k}_i$ corresponds to wave vector of incident electromagnetic wave, $\mathbf{k}_s$ corresponds to wave vector of scattered electromagnetic wave, and $\mathbf{Q}$ corresponds to the acoustic wave vector, we are observing a change of electromagnetic wave polarization. Thus, the initial linear polarization changes to the opposite linear one. For a certain direction of electromagnetic wave propagation through the TeO_2-crystal we could change the polarization of electromagnetic wave at a certain wavelength, λ, and this wavelength is directly related to the acoustic wave frequency, f. By varying f we could continuously tune the wavelength, λ, which changes the polarization, and this wavelength could be easily isolated from the polychromatic spectrum by using the polarizer.

The principle of DAOTF operation could be described as follows [29]. Initial polychromatic radiation emitted by the object of interest is passing through the linear polarizer (P_1) and reach the TeO_2-crystal of first AOF. This AOF changes the polarization of particular wavelength, λ, within the polychromatic spectrum, and this wavelength is defined by the acoustic wave frequency, f. By passing the beam through the second polarizer (P_2), which is transparent for λ polarization and almost non-transparent for polarization of the polychromatic spectrum, we are

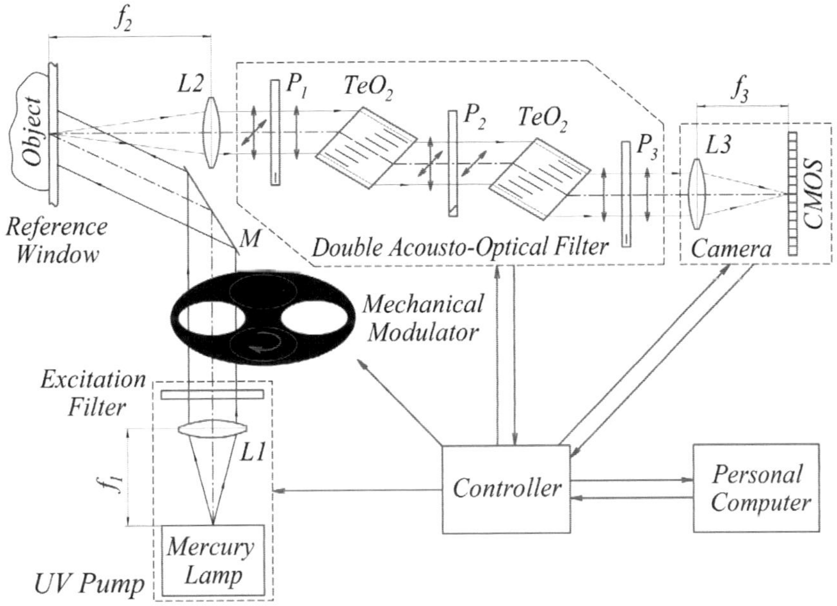

Figure 1. Schematic representation of the set-up for modulation fluorescent imaging, where M is a mirror, P_1, P_2, and P_3 are polarizers transparent for the indicated polarizations, L_1, L_2, and L_3 are lenses, collimating the pumping beam and build the object image at the plane of CMOS detector (f_1, f_2, and f_3 define focal lengths of the lenses). Wide arrows in front of polarizers and after them indicates the light polarization states.

filtering the particular wavelength, λ. Almost monochromatic beam reaches the second TeO_2-crystal, where it undergoes the polarization change for the second time. After this rotation of electromagnetic wave polarization, the optical beam passes through the third polarized (P_3), which suppresses the remaining low-power polychromatic component, leaving the λ-component untouched. Thereby, we could observe quasi-monochromatic radiation after the DAOTF.

For the particular DAOTF we are observing 2 nm narrow bandwidth of spectral transparency with the ability to continuously tune the wavelength of electromagnetic wave within the range from 430 to 780 nm [29]. The aperture of the beam passing through the DAOTF reaches $D = 8$ mm, the angular field of view reaches $2\omega = 3^\circ$ with the angular resolution of $\delta = 0.7$ radians, providing the ability for imaging the sample. The mirror (M) in the fig. 1 is used for the compactness of pumping optical system, the lenses (L_1, L_2, and L_3) are used to collimate the pumping UV beam and to built the monochromatic images at the plane of CMOS sensor (image plane). Note, the pumping system, the DAOTF and the CMOS camera is controlled via the PC. Thus, the video images of the object fluorescence in narrow bandwidth of electromagnetic spectrum could be detected utilizing the described system.

Let us define the monochromatic video sequence with $s_\lambda(x, y, t)$ (x and y represent spatial coordinates at the image plane, t represents the time of image registration, and index λ defines the central wavelength of DAOTF transparency). The object fluorescence could be describe as a spatial distribution of the fluorescence intensity, $p_\lambda(x, y)$, and we are estimating the latter one utilizing the following procedure

$$p_\lambda(x, y) = s_\lambda(x, y, \nu = \nu_0), \tag{3}$$

where

$$\widetilde{s}_\lambda(x, y, \nu) = \int_{-\infty}^{+\infty} s_\lambda(x, y, t)\, e^{-i2\pi\nu t} dt$$

stands for Fourier transform applied to time-domain data at each point of the image plane. Equation (3) helps us to both prevent an impact of non-fluorescent background on $p_\lambda(x, y)$ reconstruction, and to suppress the digital noises inherent to single image of the video sequence, $s_\lambda(x, y, t = t')$, where t' corresponds to the certain moment of time.

3. Results

We are implementing the described experimental set-up for studying the test sample in order to verify the proposed method of modulation fluorescent imaging. We are studying the samples of polymer fluorescent thread, containing fluorescent dye and injected into a peace of white low-fluorescent paper. During the study we are setting the spectral band of DAOTF transparency to 612 nm, and we are utilizing maximal power of the UV pump. Figure 2 shows schematic representation of video sequence of fluorescent images of test object: (a) shows the theoretical time-domain profile of the UV pump intensity, $I(t)$ (equation (1)), and (b) shows the separate images of the video sequence corresponding to different pumping conditions (1st, 2nd, and 3rd fluorescent images in the figure 2 (b) correspond to (1) maximal, (2) medium, and (3) minimal power of UV pump, respectively). One could observe strong difference between the object fluorescence intensity at this array of images.

By applying the procedure (3) we are reconstructing the parametric image of object fluorescence, $p_\lambda(x, y)$, and figure 3 shows the results of these calculations: (a) corresponds to simple visual image of object corresponding to the case of no UV pump is applied, and (b) presents the result of modulation fluorescent imaging implementation (equation (3)). We observe strong suppression of digital noises, as well as suppression of non-fluorescent background in the experimental parametric image in comparison to the ordinary images of the video data.

The described technique (equation (3)) helps to significantly suppress the digital noises in the fluorescent parametric image, $p_\lambda(x, y)$, and an approximate increase of fluorescent data SNR

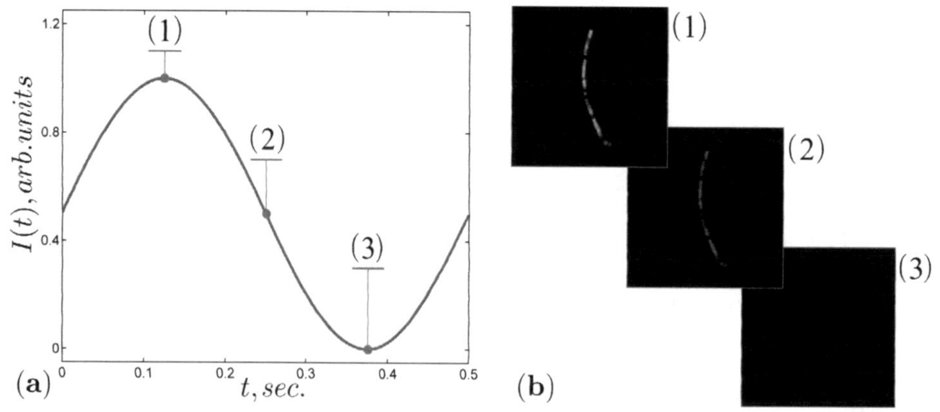

Figure 2. Video sequence of fluorescent images of test object: (a) represents theoretical time-domain profile of the UV pump intensity, $I(t)$ (equation (1)), and (b) shows the separate images of the video sequence corresponding to different pumping conditions; 1st, 2nd, and 3rd fluorescent images in the Fig. 2 (b) correspond to (1) maximal, (2) medium, and (3) minimal power of UV pump.

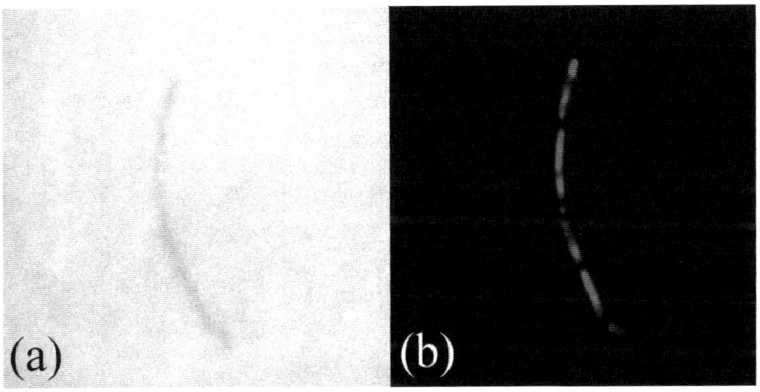

Figure 3. Modulation fluorescent imaging result: (a) corresponds to simple visual image of object with no UV pumping, and (b) presents the result of modulation fluorescent imaging implementation (equation (3).

could be described as

$$K = \frac{P_{\Delta\nu}}{P_{\nu_0}} = \nu_{frame} T, \qquad (4)$$

where $P_{\Delta\nu} \propto \nu_{frame}$ stands for the power of the digital noise localized in the entire Fourier-domain, $P_{\nu_0} \propto 1/T$ stands for the power of digital noise localized near the modulation frequency, ν_0, in the Fourier-domain, and T corresponds to the duration of entire video capture. According to equation (4), the higher the duration of video sequence we are detecting, the higher resolution in the discrete Fourier-domain we are achieving, and the lower the power of digital noises localized within the elementary discrete Fourier-domain window, centered near ν_0, we are observing. For example, for a given experiment conditions, $\nu_{frame} = 25.0$ Hz and $T = 4.0$ sec., we could achieve $K = 10^2$ times enhancement of SNR by using the modulation fluorescent imaging in comparison to the ordinary one. This estimation is in compleat accordance with the results of experimental implementation of the algorithm.

The mane advantage of the describe set-up for modulation fluorescent imaging is associated with the ability to continuously tune the central wavelength of the DAOTF transparency, λ, within the wide operating range from 430 to 780 nm. Thus, a set of N fluorescent parametric images of the sample

$$p_{\lambda_1}(x,y), p_{\lambda_2}(x,y), p_{\lambda_3}(x,y), ..., p_{\lambda_N}(x,y), \qquad (5)$$

with small spectral step $\Delta\lambda = \lambda_{i+1} - \lambda_i$ form an effective basis for further inverse problem solutions, object detection, recognition, and identification. A list of potential applications for the described technique, which we are going to consider, includes imaging and microscopy of biological tissue and cells, medical diagnosis, for instance, diagnosis of pigment skin neoplasms and skin cancers [30].

4. Conclusion

In this paper we have proposed method for multi-spectral fluorescent imaging based on acousto-optical filtering. We have used double TeO_2-based acousto-optical filter monochromator in order to construct the acousto-optical filter, tunable in range from 430 to 780 nm and having 2 nm bandwidth of spectral transparency. We have utilize modulation of UV pump intensity to reduce an impact of non-fluorescent background, as well as to suppress the digital noises in fluorescent images via the digital processing of modulated video sequence. We have studied the test sample in order to implement the proposed technique and to study the increase of signal-to-noise ratio in the fluorescent parametric image.

Acknowledgments

This work was supported by the grant of Russian Scientific Foundation, Project # 14-15-00758.

References

[1] Yanagida T, Nakase M, Nishiyama K, Oosawa F 1984 *Nature* **307** 58–60
[2] Kron S J, Spudich J A 1986 *Proc. the National Academy of Sciences of the United States of America* **83** 6272–76
[3] Harada Y, Noguchi A, Kishino A, Yanagida T 1987 *Nature* **326** 805–8
[4] Toyoshima Y Y, Kron S J, McNally E M, Niebling K R, Toyoshima C, Spudich J A 1987 *Nature* **328** 536–39
[5] Small J V 1988 *Nature* **331** 568–69
[6] Nishizaka T, Yagi T, Tanaka Y, Ishiwata S 1993 *Nature* **361** 269–71
[7] Funatsu T, Harada Y, Tokunaga M, Salto K, Yanagida T 1995 *Nature* **374** 555–59
[8] Rice B W, Cable M D, Nelson M B 2001 *J. Biomed. Opt.* **6** 432-40
[9] Suwansukho K, Sumriddetchkajorn S, Buranasiri P 2014 *Appl. Opt.* **53** 2206–12
[10] Zeng Y, Yan B, Sun Q, He S, Jiang J, Wen Z, Qu J Y 2014 *Biomed. Opt. Exp.* **5** 653–63
[11] Cheng S, Cuenca R M, Liu B, Malik B H, Jabbour J M, Maitland K C, Wright J, Cheng Y-S L, Jo J A 2014 *Biomed. Opt. Exp.* **5** 921–31
[12] Lee J H, Kim H K, Chandhanayingyong C, Lee F Y-I, Hielscher A H 2014 *Biomed. Opt. Exp.* **5** 2301–16
[13] Tsai E H R, Bentz B Z, Chelvam V, Gaind V, Webb K J, Low P S 2014 *Biomed. Opt. Exp.* **5** 2662-78
[14] Dobosz M, Strobel S, Stubenrauch K-G, Osl F, Scheuer W 2014 *J. Biomed. Opt.* **19** 016022
[15] Fortian A, Sorkin A 2014 *J. Cell Science* **127** 432-44
[16] Ghielmetti G, Aegerter C M 2014 *Opt. Exp.* **22** 1981–89
[17] Hess S T, Girirajan T P K, Mason M D 2006 *Biophys. J.* **91** 4258-72
[18] Betzig E, Patterson G H, Sougrat R, Lindwasser O W, Olenych S, Bonifacino J S, Davidson M W, Lippincott-Schwartz J, Hess H F 2006 *Science* **313** 1642–45
[19] O' Holleran K, Shaw M 2014 *Biomed. Opt. Exp.* **5** 2580–90
[20] Favreau P, Hernandez C, Lindsey A S, Alvarez D F, Rich T, Prabhat P, Leavesley S J 2014*J. Biomed. Opt.* **19** 011017
[21] Krieger J, Singh A P, Garbe C S, Wohland T, Langowski J 2014 *Opt. Exp.* **22** 2358–75
[22] Wong Y, Chao J, Lin Z, Ober R J 2014 *Opt. Exp.* **22** 20396–420
[23] Milstein A B, Oh S, Webb K J, Bouman C A, Zhang Q, Boas D A, Millane R P 2003 *Appl. Opt.* **42** 3081–94
[24] He W, Pu H, Zhang G, Cao X, Zhang B, Liu F, Luo J, Bai J 2014 *Appl. Opt.* **53** 402–9
[25] He W, Zhang G, Pu H, Liu F, Cao X, Luo J, Bai J 2014 *J. Biomed. Opt.* **19** 056012
[26] Petrovsky A, Schellenberger E, Josephson L, Weissleder R, Bogdanov A 2003 *Cancer Research* **63** 1936-42
[27] Chen X, Conti P S, Moats R A 2004 *Cancer Research* **64** 8009-14
[28] Pozhar V E, Pustovoit V I, Kutuza I B, Perchik A V, Mazur M M, Shorin V N 2005 *Proc. SPIE* **5953** 200–3
[29] Perchik A V 2014 *Proc. SPIE* **9129** 91293A
[30] Vogan K 2008 *Nature Reviews Genetics* **9** 502

Effects of carrier-carrier scattering on population inversion in graphene under pulse photoexcitation

Akira Satou, Victor Ryzhii and Taiichi Otsuji

Tohoku University, Research Institute of Electrical Communication, 2-1-1 Katahira, Aoba-ku, Sendai 980-8577, Japan

E-mail: a-satou@riec.tohoku.ac.jp

Abstract. We study the carrier relaxation dynamics in intrinsic graphene after pulse photoexcitation and reveal effects of intraband carrier-carrier scattering on population inversion in the terahertz region, by conducting simulation based on the quasi-classical Boltzmann equation. It is demonstrated that by changing the dielectric constant of the surrounding materials the rate of carrier-carrier scattering can be controlled and the relaxation dynamics differs for cases with low and high dielectric constants. It is also found that the Pauli blocking of photogeneration in case of the pulse photoexcitation causes decrease in the photocarrier concentration and thus weakening of population inversion with higher dielectric constant.

1. Introduction

Device applications of graphene utilizing its unique electronic/optical properties have been investigated extensively. In particular, We have studied terahertz (THz) lasers and active THz plasmonic devices [1-3] based on interband population inversion in graphene at room temperature, which take advantage of the high carrier mobility and gapless linear dispersion of graphene. The zero bandgap enables population inversion in very low energy region, i.e., down to the THz region. Several pumping schemes such as optical pumping and current injection have been proposed [2-5]. Experimental observations of light amplification by stimulation emission from graphene under pulse photoexcitation have already been reported in the THz [7] and near-infrared [6] regions.

For any pumping schemes, the carrier-carrier (CC) scattering is anticipated to play a crucial role in the achievement of population inversion. Very fast quasi-equilibration of carriers by the CC scattering in comparison with cooling by the optical-phonon (OP) emission hinders population inversion with optical pumping [8, 9]. The rise of the carrier temperature due to the quasi-equilibrium causes less accumulation of carriers in the the THz region. It has been shown in Ref. [8] that, if the time scale of cooling due to the OP emission is faster than the quasi-equilibration, the threshold pumping intensity for the population inversion is much smaller than the case of the fast quasi-equilibration.

The characteristic time scale of the CC scattering can be changed, relatively to other mechanisms such as OP scattering, via the choice of surrounding dielectric materials ($\tau_{cc} \propto \epsilon^2$) and/or via the presence of gate electrodes. They can be utilized for either suppressing or enhancing the effect of the CC scattering and, in turn, for optimizing device performances. To fully reveal the role of the CC scattering on operation principles and its controllability via

Content from this work may be used under the terms of the Creative Commons Attribution 3.0 licence. Any further distribution of this work must maintain attribution to the author(s) and the title of the work, journal citation and DOI.
Published under licence by IOP Publishing Ltd

dielectric materials and/or gate electrodes, numerical simulation taking into account the CC scattering is inevitable.

In this paper, we develop a kinetic transport model with the CC scattering based on a deterministic solution approach for the semi-classical Boltzmann equation and on the direct evaluation of the collision integral for the CC scattering. Using the model, we conduct transient simulation of carrier relaxation dynamics in intrinsic graphene after femtosecond photoexcitation. We demonstrate that with high dielectric constant of the surrounding materials the suppression of CC scattering induces not only cooling by OP emission but also the Pauli blocking of photogeneration of carriers, so that the threshold of pulse fluence for population inversion becomes larger with higher dielectric constant in case of pulse photoexcitation.

2. Simulation model

For a first step towards the development of a device simulation model for graphene THz devices with inclusion of the CC scattering, we start from a spatially homogeneous case (its extension to the inhomogeneous case is straightforward, although the former is computationally more expensive). Our model is based on semi-classical Boltzmann equations for electrons and holes:

$$\frac{\partial f_{c\boldsymbol{p}}}{\partial t} = \sum_{c'=e,h} J_{cc'} + \sum_{i} J_{ci}, \tag{1}$$

where $f_{c\boldsymbol{p}}$ is the distribution function for carrier type c (e and h for electrons and holes, respectively). Here, $J_{cc'}$ is the collision integral for c-c' scattering:

$$\begin{aligned} J_{cc'}(\boldsymbol{p}) = {} & \frac{4}{(2\pi\hbar)^4} \int d^2\boldsymbol{p}_1 \int d^2\boldsymbol{p}' W_{cc'}(\boldsymbol{p}, \boldsymbol{p}_1; \boldsymbol{p}', \boldsymbol{p}'_1) \\ & \times \left[f_{c\boldsymbol{p}'} f_{c'\boldsymbol{p}'_1} (1 - f_{c\boldsymbol{p}})(1 - f_{c'\boldsymbol{p}_1}) - f_{c\boldsymbol{p}} f_{c'\boldsymbol{p}_1} (1 - f_{c\boldsymbol{p}'})(1 - f_{c'\boldsymbol{p}'_1}) \right] \end{aligned} \tag{2}$$

with $\boldsymbol{p}'_1 = \boldsymbol{p} + \boldsymbol{p}_1 - \boldsymbol{p}'$ and $W_{cc'}$ being the Coulomb scattering probability. $W_{cc'}$ contains a delta function representing the energy conservation, $\delta(p + p_1 - p' - p'_1)$, so that Eq. (2) reduces to a triple integral. For simplicity, we consider the static Thomas-Fermi screening for the Coulomb potential (effects of the dynamic screening shall be briefly discussed at the end of the next section).

For other scattering mechanisms which are represented by J_{ci}, we consider the intraband and interband OP emission/absorption, $J_{c,\mathrm{op}}$, as well as photogeneration, $J_{c,\mathrm{ph}}$. Here, we assume equilibrium optical phonons. For the photogeneration, we assume a Gaussian pulse with central wavelength λ, pulse width Δt (full-width at half-maximum), and pulse fluence J.

We developed a computer program that solves Eq. (1), first by evaluating $J_{cc'}$ at each point (p_i, θ_j) in parallel and then by applying the third-order TVD Runge-Kutta scheme [10] for the time-step update.

3. Results and discussions

We conducted transient simulation of carrier relaxation dynamics in intrinsic graphene after femtosecond photoexcitation. As parameters for the Gaussian pulse we fixed $\lambda = 1.55$ μm and $\Delta t = 80$ fs, and we set the initial distribution functions of electrons and holes with temperature $T = 300$ K and zero Fermi level (i.e., intrinsic graphene). We varied the dielectric constant of the surrounding material, ϵ, to control the time scale of the CC scattering and thus to reveal its effect on population inversion.

Figure 1 shows snapshots of the electron distribution function with relatively low dielectric constant $\epsilon = (\epsilon_{\mathrm{air}} + \epsilon_{\mathrm{SiO_2}})/2 = 2.75$ and the pulse fluence $J = 6$ μJ/cm^2. The angle θ is taken

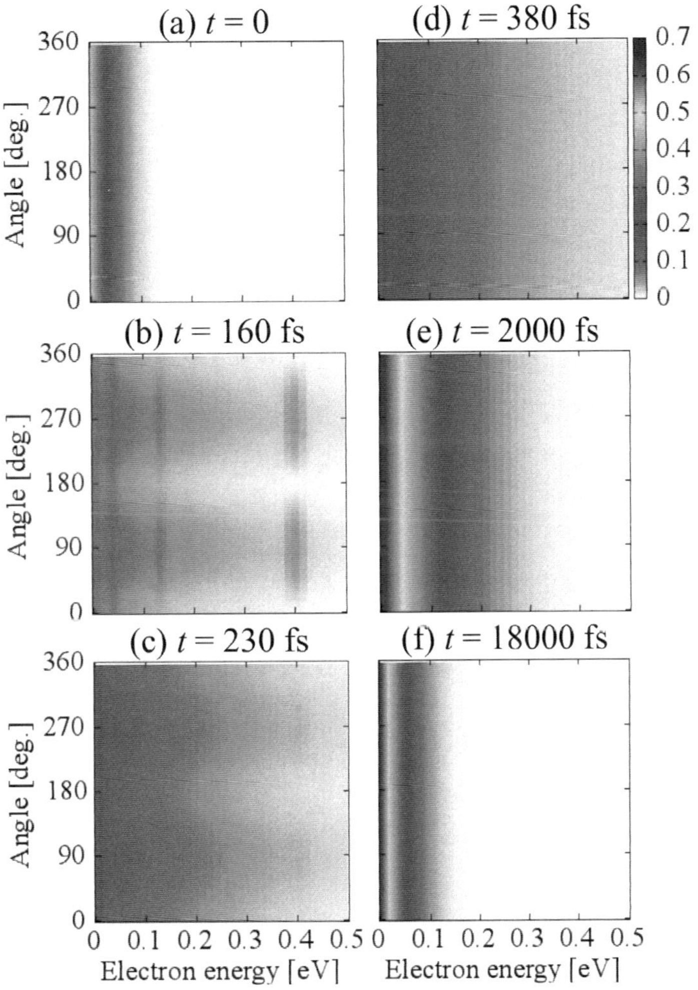

Figure 1. Snapshots of the electron distribution function with $\epsilon = 2.75$ and $J = 6\ \mu J/cm^2$ that show distinct states of relaxation dynamics: (a) initial distribution, (b) photoexcitation at the peak of the optical pulse, (c) quasi-equilibration in the energy direction, (d) quasi-equilibration in the angular direction, (e) cooling by OP emission and population inversion in the THz region (< 40 meV), and (f) relaxation of population inversion by recombination due to OP emission. Note that the condition of population inversion is $f > 1/2$.

with respect to the polarization of the pulse. Figure 1 clearly illustrates distinct states of carrier relaxation dynamics after the pulse photoexcitation; the photogenerated carriers together with the initial carriers are quasi-equilibrated in the energy direction in the phase space within a few hundred fs (Fig. 1(c)), followed by quasi-equilibration in the angular direction (d), then they are cooled down by OP emission after a few ps and the population inversion in the THz region is achieved (e), and the recombination takes place in a few tens of ps (f). Note that the condition of population inversion for intrinsic graphene is $f > 1/2$. The difference between time scales of quasi-equilibration in energy and angular directions originates from the small-angle nature of the Coulomb scattering.

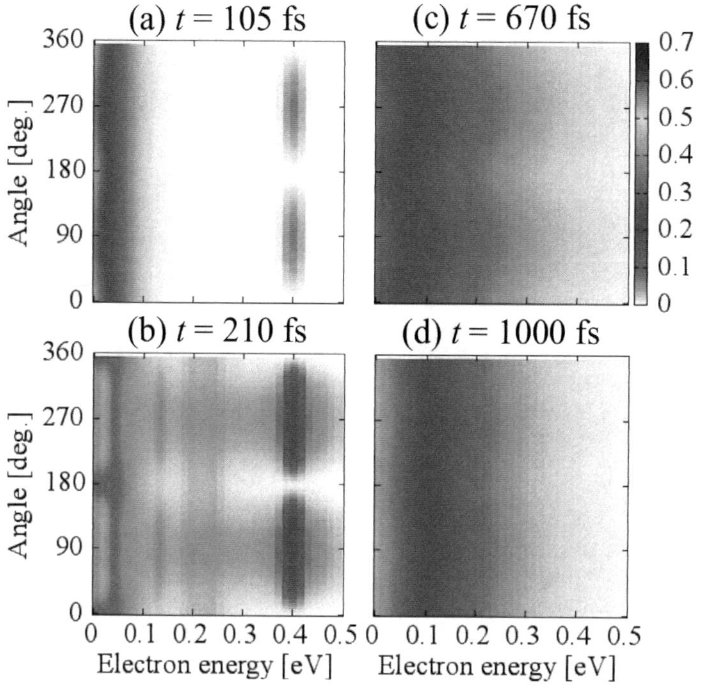

Figure 2. Snapshots of the electron distribution function with $\epsilon = 40$ and $J = 8\ \mu\mathrm{J/cm^2}$: (a) photoexcitation before the peak of the optical pulse, (b) photoexcitation after the peak of the optical pulse, exhibiting a wide peak around $\varepsilon = 0.2 - 0.24$ eV associated with single OP emission, (c) quasi-equilibration in the energy direction, (d) partial quasi-equilibration in the angular direction.

The time dependence of distribution in Fig. 1 after quasi-equilibration in both directions is almost identical with that calculated in Ref. [9], which is based on the assumption that the distribution is always quasi-equilibrium even during the photoexcitation. This is because the time scale of the OP emission (a few hundred fs) is slower than the completion time of quasi-equilibration, with the low value of the dielectric constant, $\epsilon = 2.75$.

The picture of carrier relaxation dynamics changes when the dielectric constant is high and the CC scattering is suppressed. Figure 2 shows snapshots of the electron distribution function for very high dielectric constant $\epsilon = 40$ and the pulse fluence $J = 8\ \mu\mathrm{J/cm^2}$, before the completion of quasi-equilibration. In Fig. 2(b), it is seen that a small, wide peak around $\varepsilon = 0.2 - 0.24$ eV appears. The energy difference of this peak from the photogeration peak ($\varepsilon = 0.4$ eV) corresponds to the energies of optical phonons (0.16 eV for K optical phonons and 0.2 eV for Γ optical phonons), indicating that it originates from single OP emission by photogenerated carriers. It is also worth mentioning that quasi-equilibration in the angular direction is done both by CC scattering and OP emission (Fig. 2(c)), although a small deviation of the distribution function by angle is kept for a long time in the low energy region (over 1 ps, see Fig. 2(d)) due to the prohibition of OP emission by energy conservation low.

Figure 2(b) also shows another distinct feature; a very large value of the photogeneration peak even after the peak of Gaussian pulse. This means that the large fraction of photogenerated carriers stay longer at the same energy than the time scale of the photogeneration (80 fs). It causes the Pauli blocking of photogeneration and, in turn, the decrease in the photocarrier

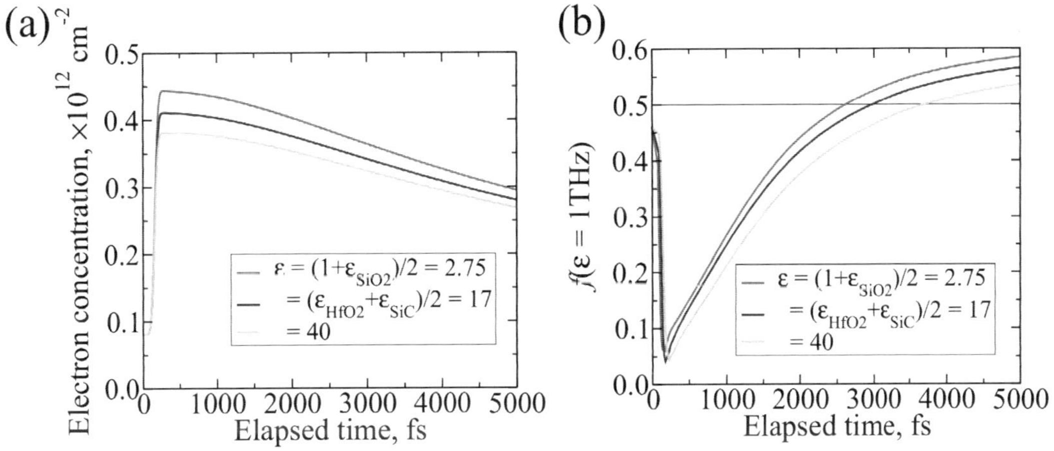

Figure 3. (a) Time evolution of the total electron concentration after the pulse photoexcitation with $J = 2\ \mu\mathrm{J/cm}^2$ and with different dielectric constants. (b) The time evolution of electron distribution function at $\varepsilon = 2.07$ meV corresponding to half of the photon energy with frequency 1 THz.

concentration (see Fig. 3(a)). As shown in Fig. 3(b), the effect of Pauli blocking on population inversion is stronger than that of cooling by OP emission, so that the population inversion achieved by a fixed pulse fluence becomes weaker for higher dielectric constant.

Contrary to the case of pulse excitation studied here, however, the lowering of the threshold pumping intensity with higher dielectric constant is still expected in the case of CW pumping, as the effect of Pauli blocking is negligibly small for the pumping intensity necessary for the THz population inversion. Besides, the static screening of the Coulomb potential assumed in this work overestimates the time scale of quasi-equilibration by the CC scattering, as it ignores the dynamic screening for the high-energy photogenerated carriers. Especially, it is pointed out that the CC scattering with small-angle change is suppressed by the dynamic screening [11]. It should result in the further suppression of carrier heating and thus lowering of the threshold. For quantitative analysis of this, it is very crucial to include the dynamic screening in the simulation model.

4. Conclusion

We developed a simulation model for graphene devices taking into account the CC scattering. Based on the model, we conducted a transient simulation of carrier relaxation dynamics in intrinsic graphene after femtosecond photoexcitation. It was revealed that the relaxation dynamics can be changed by the dielectric constant of the surrounding materials. With low dielectric constant, quasi-equilibration takes place much faster than cooling by OP emission. With sufficiently high dielectric constant, on the other hand, we observed a peak in the distribution function before quasi-equilibration, which is attributed to single OP emission from the photogeneration peak. Simultaneously, it was found that the Pauli blocking of photogeneration in case of the pulse photoexcitation weakens population inversion with higher dielectric constant.

Acknowledgment

This work was financially supported by JSPS Grant-in-Aid for Young Researcher (B) (#23760300) and by JSPS Grant-in-Aid for Specially Promoted Research (#23000008). This work used computational resources of the HPCI systems provided by Research Institute for Information Technology, Kyushu University, through the HPCI System Research Project (Project ID:hp120143), and by Information Technology Center, Nagoya University (Project ID:hp140086).

References

[1] Otsuji T, Boubanga Tombet S A, Satou A, Fukidome H, Suemitsu M, Sano E, Popov V, Ryzhii M and Ryzhii V 2012 *J. Phys. D: Appl. Phys.* **45** 303001
[2] Ryzhii V, Ryzhii M and Otsuji T 2007 *J. Appl. Phys.* **101** 083114
[3] Ryzhii M and Ryzhii V 2007 *Jpn. J. Appl. Phys.* **46** L151
[4] Davoyan A R, Morozov M Yu, Popov V V, Satou A and Otsuji T 2013 *Appl. Phys. Lett.* **103** 251102
[5] Satou A, Vasko F T, Otsuji T and Mitin V V 2014 *J. Phys. D: Appl. Phys.* **47** 055103
[6] Li T, Luo L, Hupalo M, Zhang J, Tringides M C, Schmalian J and Wang J 2012 *Phys. Rev. Lett.* **108** 167401
[7] Boubanga-Tombet S, Chan S, Watanabe T, Satou A, Ryzhii V and Otsuji T 2012 *Phys. Rev. B* **85** 035443
[8] Ryzhii V, Ryzhii M, Mitin V, Satou A and Otsuji T 2011 *Jpn. J. Appl. Phys.* **50** 094001
[9] Satou A, Ryzhii V, Kurita Y and Otsuji T 2013 *J. Appl. Phys.* **14** 143108
[10] Shu C-W and Osher S 1988 *J. Comp. Phys.* **77** 439
[11] Brida D *et al* 2013 *Nature Communications* **4** 1987

Journal of Physics: Conference Series **584** (2015) 012019 doi:10.1088/1742-6596/584/1/012019

Scattering in structured two-layered medium

Irina N Dolganova*, Aleksandra S Neganova, Valeriy E Karasik, Kirill I Zaytsev and Stanislav O Yurchenko

Bauman Moscow State Technical University, 2nd Baumanskaya str. 5, 105005, Moscow, Russia

E-mail: *in.dolganova@gmail.com

Abstract. Multiple scattering in two-layered medium was modeled using Monte Carlo numerical simulation. Special case of group formation in such layers was considered. To estimate the impact of structure factor on the overall results, the rigorous calculation of single scattering phase function was used. Comparison with the Monte Carlo simulations for the respective model phase function shows that there were some differences in the scattered radiance. These differences had no significant energy redistribution in the angular dependence of the scattered radiance, but different elongation in the zero-direction occurred. So it was shown that the group formation and special structure still leaded to some changes in the scattering process, even if the medium had several layers with different parameters and had rather large width.

1. Introduction

One of the most difficult cases in the scattering problem is the wave propagation through a medium with densely packed particles, like colloids, powders, photonic crystals, etc. [1], [2], [3], [4], [5], [6], [7], [8], [9]. Most approximations and strong solutions do not work for densely package, because particles are no more independent and thus multiple scattering should be taken into account [10], [11], [12]. The order of particles in such medium can be random or periodic. Also particles can be joined in groups distributed randomly in the scattering volume. This group formation characterizes such media as clouds with ice crystals, liquids at an initial stage of crystallization, disperse systems during the coagulation process. It is evident, that the dense package leads to interference effects for scattering on one group of particles [8], [13]. But the random character of groups distribution may remove such effects. And respectively, the radiation transfer will have no sensitivity to them. As the result, the random approximation for scattering medium is often used.

Monte Carlo (MC) numerical modeling [14], [15], [16], [17], [18] is an effective way to solve radiation transfer equation (RTE) [19]:

$$L(x) = L_0(x) + \int k(x,x')L(x')dx', \tag{1}$$

where the integral kernel is:

$$k(x,x') = \frac{\Lambda\varepsilon}{4\pi}\chi(\hat{\boldsymbol{I}},\hat{\boldsymbol{I}}')\delta\left(\frac{\boldsymbol{r}-\boldsymbol{r}'}{|\boldsymbol{r}-\boldsymbol{r}'|}\right)\frac{\exp(-\varepsilon|\boldsymbol{r}-\boldsymbol{r}'|)}{(\boldsymbol{r}-\boldsymbol{r}')^2}. \tag{2}$$

where $L_0(\boldsymbol{r},\hat{\boldsymbol{I}})$ is radiance of incident field at the point $x = (\boldsymbol{r},\hat{\boldsymbol{I}})$, Λ is single scattering albedo, ε is extinction, $\chi(\hat{\boldsymbol{I}},\hat{\boldsymbol{I}}')$ is single scattering phase function.

Content from this work may be used under the terms of the Creative Commons Attribution 3.0 licence. Any further distribution of this work must maintain attribution to the author(s) and the title of the work, journal citation and DOI.
Published under licence by IOP Publishing Ltd

Phase function $\chi(\hat{\boldsymbol{I}}, \hat{\boldsymbol{I}}')$ characterizes the interaction between the incident radiation and a single scattering group. In order to account the wave character of scattering process, it should be determined as precise as possible. It is possible to use the numerical solution of Maxwell equations for this purpose [20], [21], [22], [23]. In order to reduce computational time, we will deal only with 2D model of particles and consider that they have cylindrical form. As it was shown in [24], for different particle groups single scattering phase function significantly depends on the group structure type. Also in [24] it was shown, that structure factor has an essential influence on the angular distribution of scattered radiance for one scattering layer.

Here in this work we will discuss the scattering in the two-layered medium. Computational results will be obtained using Monte Carlo simulations in two cases - first, for the rigorously calculated phase functions, second, for the averaged phase function. Since the last one is often used for random medium, it will be possible to find out, whether two-layered medium may be approximated by the random layers.

2. Two-layered structured medium

The considered medium consists of two layers with different properties of scattering groups (figure 1). Scattering group from the first layer is formed by 3 identical particles oriented along OX axis with radius $R/\lambda = 0.2$, period $T/\lambda = 3$ and dielectric permittivity $\epsilon = 2$; groups from the second layer have 4 particles with the same properties, where λ is the wavelength of TEM polarized radiation, incident in OZ direction (figure 2).

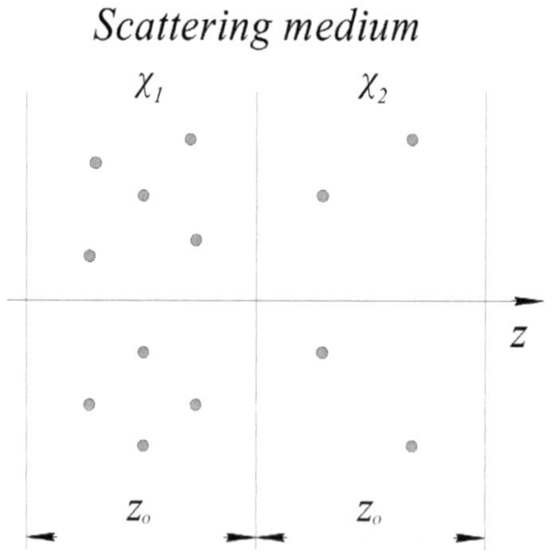

Figure 1. Scattering medium model formed by two layers.

Using FDTD numerical algorithm [20], [21], [22] for finding scattered field distribution and the Kirchhoff diffraction integral for complex field amplitudes we can calculate the single scattering phase function $\chi(\theta)$, θ is the scattering angle. It is shown in the figure 3.

Phase function is characterized by the asymmetry parameter [25]:

$$g = \int_0^\pi \chi(\theta) \sin\theta \cos\theta d\theta, \tag{3}$$

which was calculated for the obtained $\chi(\theta)$ from figure 3. Numerical values are, respectively, $g_1 = 0.2598$ for the first layer and $g_2 = 0.0061$ for the second layer.

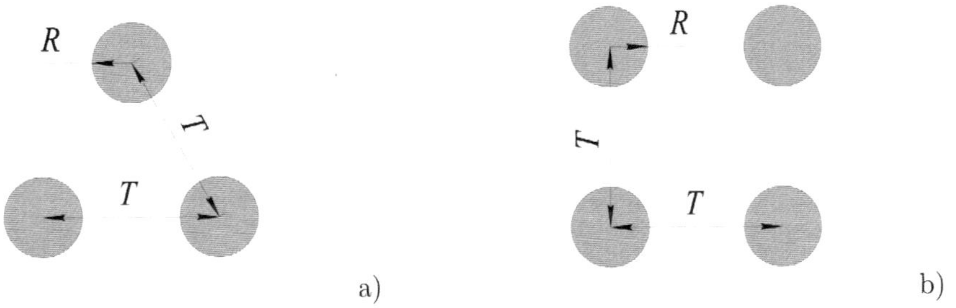

Figure 2. Form and orientation of particles in the groups of a) 3 cylinders and b) 4 cylinders (2D model).

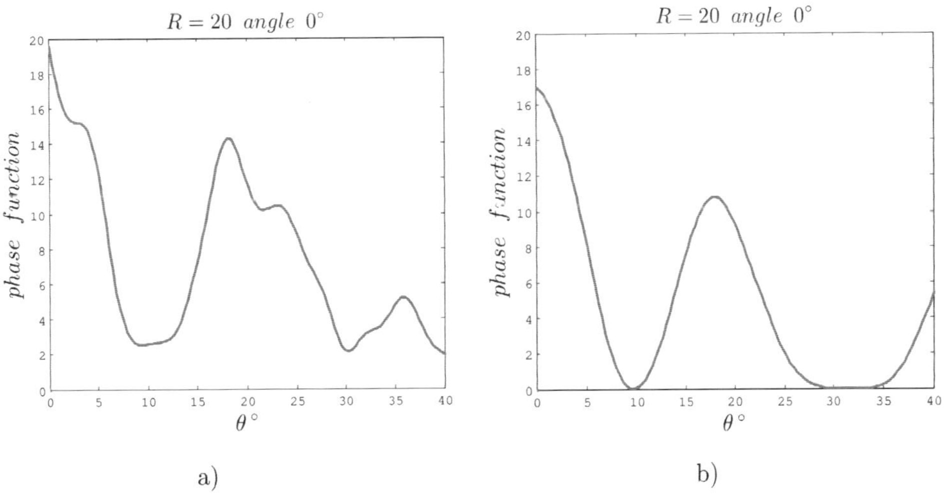

Figure 3. Scattering phase functions for groups of a) 3 particles and b) 4 particles.

Parameter g, which is the mean cosine $< \cos\theta >$, shows the degree of energy redistribution from the zero-direction to other angles during a single scattering event. It is also used for a random medium as the main parameter for a Henyey-Greenstein phase function model [26]:

$$\chi^{HG}(\theta, g) = \frac{1}{4\pi} \frac{1 - g^2}{(1 + g^2 - 2g\cos\theta)^{3/2}}. \tag{4}$$

We will use this average phase function for Monte Carlo simulations of multiple scattering in the considered two-layered medium.

3. Monte Carlo simulation

For the numerical simulation of the multiple scattering, MC computational scheme with single isotropic source was used. The scattered radiance in the single point in the zero direction can be determined applying double local estimation in MC simulation [17]. In order to do this, the scattered radiance should be divided into zero, first, and higher orders:

$$L(x) = L_0(x) + L_1(x) + \int k_I(x,x')L(x')dx', \tag{5}$$

$$k_I(x,x') = \left(\frac{\Lambda\varepsilon}{4\pi}\right)^2 \chi(\hat{\boldsymbol{I}},\hat{\boldsymbol{I}}'')\chi(\hat{\boldsymbol{I}},\hat{\boldsymbol{I}}'')\frac{\exp(-\varepsilon|\boldsymbol{r}''-\boldsymbol{r}'|)}{(\boldsymbol{r}''-\boldsymbol{r}')^2}, \tag{6}$$

the first and higher orders of scattered radiance should be found separately [24]. For the first order scattering there is the following equation [27]:

$$L_1(\theta) = \frac{\Lambda\varepsilon\exp(-\varepsilon r)}{4\pi r\sin\theta}\int_\theta^\pi \chi(\eta)\exp\left[-\varepsilon r\sin\theta\left(\tan\frac{\eta}{2}-\tan\frac{\theta}{2}\right)\right]d\eta. \tag{7}$$

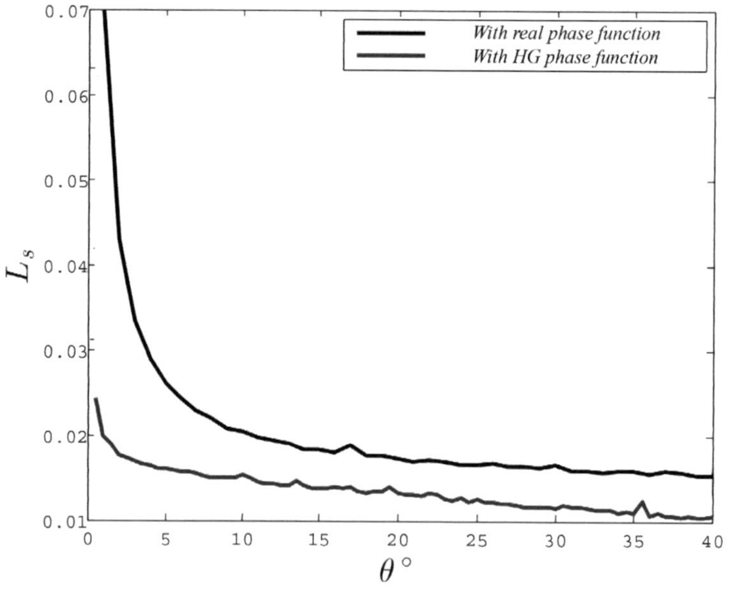

Figure 4. Scattered radiance numerically obtained for two cases: using Henyey-Greenstein phase function and using calculated ("real") phase function.

Higher order scattered radiation will be determined from the MC simulation. For both first and higher orders the probability of crossing the border between layers was equal to unity, because the layers have no real interface. According to the double local estimation in MC simulation, integration of higher orders in the RTE equation (1) over the angular space is applied in order to find the radiance at a given point $\boldsymbol{r}$ as a mean value of a random variable:

$$L_2(x) = \mathrm{M}\xi(x), \tag{8}$$

$$\xi(x) = \Sigma_{i=1}^N Q_i h_1(x_i), \tag{9}$$

$$h_1(x_i) = \chi_1(\theta_1)\chi_2(\theta_2)\frac{\exp(-\varepsilon|r-r'|)}{(r-r')^2}, \tag{10}$$

where Q_i is a "weight" of a photon after the i-th collision. If the medium single phase function is the same for different scattering events, $\chi_1(\theta_1)$ and $\chi_2(\theta_2)$ can be substituted in (10) with $\chi(\theta)$.

The number of simulations was $N = 5000$, layer length was $z_0 = 2$m. In the figure 4 the scattered radiance is shown for both computational cases (with Henyey-Greenstein and numerically obtained phase functions). As it can be noticed, differences in the obtained curves are presented, since the structure group formation still impact on the scattered radiance, comparable with the MC results for the averaged phase function, where this impact vanished. On the other hand, special redistribution of energy from zero-direction to other angles is no more present, which can be explained by the large length of layers and by the mismatch of secondary maxima for single layer scattering.

4. Conclusion

As we can see from the obtained results, there are no obvious impact of the group formation on the scattered radiation. But also there are no mismatch with the results of numerical simulations using Henyey-Greenstein phase function. All these facts show that such rather large two-layers medium behaves like a random one, but can not be approximated by random model with the single-parameter Henyey-Greenstein scattering phase function.

Acknowledgement

This work was supported by the Russian Scientific Foundation (grant # 14-19-01083).

References

[1] Budak VP, and Korkin SV 2008 *J. Quant. Spectrosc. Radiat. Transfer* **109** 1347–62
[2] Fokina IN, Zaytsev KI, Karasik VE, Tsapenko KP 2013 *Proc. SPIE Terahertz Emitters, Receivers, and Applications IV* **8846**
[3] Kokhanovsky AA 1998 *Journal of Geophysical Research: Atmospheres* **103** 6089–96
[4] Kyurkchan AG, Skorodumova EA 2009 *J. Quant. Spectrosc. Radiat. Transfer* **110** 1335–44
[5] Kirillin M, Meglinski I, Kuzmin V, Sergeeva E, and Myllylä R 2010 *Opt. Exp.* **18** 21714–24
[6] Ralston TS, Marks DL, Carney PS, and Boppart SA 2006 *J. Opt. Soc. Am. A.* **23** 1027–37
[7] Lewis P, Disney M 2007 *Remote Sensing of Environment* **109** 196–206
[8] Wolf S, Voshchinnikov NV, Henning Th 2002 *Astronomy and Astrophysics* **385** 365–76
[9] Waterman PC 1971 *Phys. Rev. D* **3** 825–939
[10] Mishchenko MI, LD Travis, AA Lacis 2006 *Multiple Scattering of Light by Particles: Radiative Transfer and Coherent Backscattering* (Cambridge: Cambridge University Press)
[11] Twersky V 1964 *Proc. Am. Math. Soc. Symp. on Stochastic Processes in Mathematical Physics and Engineering* **16** 84–166
[12] Ishimaru A 1978 *Wave Propagation and Scattering in Random Media* (London: Academic Press)
[13] Mishchenko MI, Travis LD, and Mackowski DW 1996 *J. Quant. Spectrosc. Radiat. Transfer* **55** 535–75
[14] Metropolis N, Rosenbluth AW, Rosenbluth MN, Teller AH, and Teller E 1953 *J. Chem. Phys.* **21**
[15] Sobol' IM 1973 *Monte-Carlo numerical methods* (Moscow: Nauka)
[16] Ermakov SM, Mikhailov GA 1976 *Statistical modeling* (Moscow: Nauka)
[17] Marchuk GI 1976 *Monte-Carlo method in atmosphere optics* (Moscow: Nauka)
[18] Stankevich D, Shkuratov Yu 2004 *J. Quant. Spectrosc. Radiat. Transfer* **87** 289–96
[19] Schuster A 1905 *Astrophys. J.* **21**
[20] Taflove A, Hagness SC 2000 *Computational Electrodynamics: The Finite-Difference Time-Domain Method* (Boston: Artech House)
[21] Schneider JB 2010 *Understanding the Finite-Difference Time-Domain Method* (Online book)
[22] Hao Y, Mittra R 2009 *FDTD Modeling of Metamaterials. Theory and Applications* (Boston: Arteck House)
[23] Umashankar K, Taflove A 1982 *IEEE Transac. Electromagnetic Compatibility* **EMC-24** 397–405
[24] IN Fokina, VE Karasik, VM Orlov, VP Budak 2014 *J. Quant. Spectrosc. Radiat. Transfer* **149** 108
[25] Mishchenko MI 1994 *J. Quant. Spectrosc. Radiat. Transfer* **52** 95–110
[26] LG Henyey, JL Greenstein 1941 *Astrophys. J.* **93** 70
[27] VI Savenkov, GN Mel'nikov 1979 *Light fields in ocean* (Moscow: AS USSR)

The absolute calibration of high-precision optical flats across a wide range of spatial frequencies

Yana V Gladysheva[1], **Ilya V Zhivotovsky**[1], **Dmitry G Denisov**[1], **Nikolay V Baryshnikov**[1], **Valery E Karasik**[1] **and Paul Rees**[2]
[1] Bauman Moscow State Technical University, 2nd Baumanskaya str. 5, 105005, Moscow, Russia
[2] Glyndŵr University, St Asaph Business Park, St Asaph, LL17 0JD, North Wales, UK

E-mail: yanagladysheva@gmail.com

Abstract. We analyse the applicability of the two-flat-test calibration method across wide spatial frequency range based on its calculated accuracy. A number of simulations have been performed to determine the accuracy of the absolute calibration. The simulation results show that the two-flat-test calibration method is applicable to surfaces within the spatial frequency range from 1.67×10^{-1} to 1.67×10^{-3} mm^{-1}.

1. Introduction

The surface form of large optical flats is critical to the performance of powerful laser systems and optical telescopes. The measurement of the surface topography and optical aberrations of these optics with nanometre resolution is a challenging problem [1-3]. Usually, the Fizeau interferometer [4] is used to measure the surface accuracy of optical flats and the result of this measurement is a difference between the test and the reference flat. In the case of highly precise optics, the topography of the reference flat must be established through absolute calibration methods and this calibration is subsequently used to accurately quantify the surface topography of the test flat.

Early absolute calibration methods for optical flats used liquid mirrors [5-8], but these methods have problems associated with vibration and the capillary effect near the wall of the vessel that holds the liquid. Moreover, mirrors must be calibrated in horizontal orientation, which results in a sag error caused by gravity and which must be compensated for in some way. Later, a three-flat-test method was developed [9-12], which compares three flat surfaces using the Fizeau interferometer. This method determines only a linear profile of the test flat along one diameter. To obtain the three-dimensional topography of the test flat, extended methods of the three-flat-test have been proposed [13-30]. These methods involve additional measurements with rotation or translation of at least one surface. The most widely used methods are based on Zernike polynomials [13-17], or even and odd functions as described by Ai and Wyant [18-23], or rotation and mirror symmetry [24-28]. Methods with a horizontally mounted reference flat are also used [29, 30]. However, none of these methods enable the measurement of the topography of large aperture optical components across a wide spatial frequency range (e.g. from 4×10^{-1} to 1.67×10^{-3} mm^{-1}) with desired accuracy.

In this study we analyse the applicability of the two-flat-test calibration method (designed for French project Laser Megajoule [31]) over a wide spatial frequency range based on its calculated accuracy. Numerical simulations of the surface reconstruction utilizing surface translations in horizontal (x) and vertical (y) directions are used to compute the accuracy of the method. These

Content from this work may be used under the terms of the Creative Commons Attribution 3.0 licence. Any further distribution of this work must maintain attribution to the author(s) and the title of the work, journal citation and DOI.
Published under licence by IOP Publishing Ltd

simulation results show that the two-flat-test calibration method is applicable to surfaces within a spatial frequency range from 1.67×10^{-1} to 1.67×10^{-3} mm^{-1}.

The remainder of this paper is organized as follows: in Section 2 we introduce the general aspects of the two-flat-test calibration method in application to our experimental setup; in Section 3 simulation results are presented of the two-flat-test calibration method for the specific spatial frequency ranges of interest; and our conclusions are presented in Section 4.

2. Method description

The two-flat-test calibration method was used with a Fizeau interferometer plus an additional collimator in the form of a spherical mirror. The spherical mirror was used to illuminate an aperture of a large optical flat of about 1000mm diameter. A reference flat was designed as an optical wedge to reflect the beam on the spherical mirror. A simplified layout of the experimental setup is presented in Figure 1.

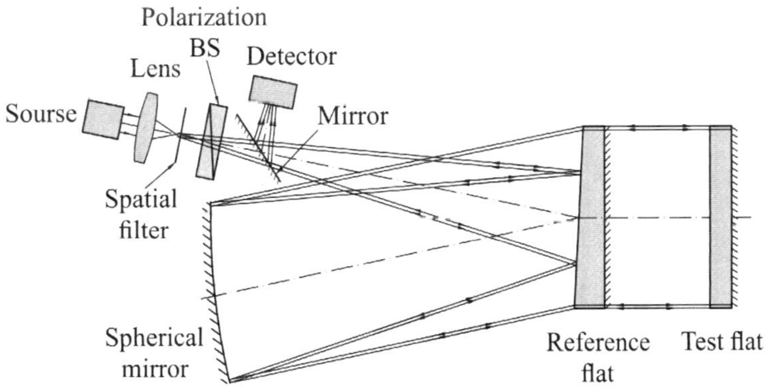

Figure 1. Layout of the large aperture interferometer, showing major components: light source; spherical mirror, extending the beam; reference flat, performed by the optical wedge; and the test flat.

The two-flat-test calibration method is based on the translation of the test flat with respect to the reference flat on a distance T along the vertical and horizontal axes (see Figure 2). The measurements may be represented mathematically in matrix form (following Küchel [24]):

$$\begin{bmatrix} m_1(x,y) \\ m_2(x,y) \\ m_3(x,y) \\ m_4(x,y) \end{bmatrix} = \begin{bmatrix} 1 & 1 & 0 & 0 & 0 \\ 1 & 0 & 1 & 0 & 0 \\ 1 & 0 & 0 & 1 & 0 \\ 1 & 0 & 0 & 0 & 1 \end{bmatrix} \cdot \begin{bmatrix} h_{reference}(-x,y) \\ h_{test}(x+T/2,y) \\ h_{test}(x-T/2,y) \\ h_{test}(x,y+T/2) \\ h_{test}(x,y-T/2) \end{bmatrix}, \quad (1)$$

where $h_{ref}(-x,y)$ and $h_{test}(x,y)$ are the surface height distributions of the reference and test flats, respectively.

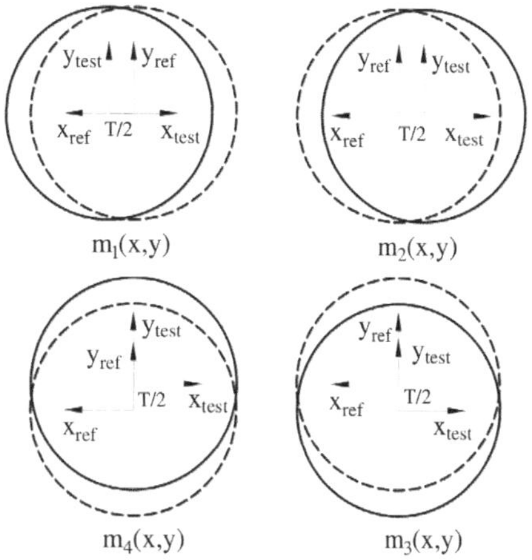

Figure 2. Four configurations of the two-flat-test method. The test flat (solid line) moves with respect to the reference flat (dashed line) along horizontal axis over T/2 left ($m_1(x,y)$) and over T/2 right ($m_2(x,y)$) along vertical axis over T/2 up ($m_3(x,y)$) and over T/2 down ($m_4(x,y)$).

The surface height distribution of the test flat $h_{test}(x,y)$ is computed as follows:

$$h_{test}(x,y) = \\ = \iint \left(\frac{1}{2i\pi(\nu_x^2+\nu_y^2)} \left(\begin{array}{l} \nu_x FT\left[\iint 2i\pi\nu_x FT(h_{test}(\nu_x,\nu_y))e^{2i\pi(\nu_x x+\nu_y y)} d\nu_x d\nu_y\right] + \\ +\nu_y FT\left[\iint 2i\pi\nu_y FT(h_{test}(\nu_x,\nu_y))e^{2i\pi(\nu_x x+\nu_y y)} d\nu_x d\nu_y\right] \end{array} \right) \right) e^{2i\pi(\nu_x x+\nu_y y)} d\nu_x d\nu_y, \quad (2)$$

where $FT[\ ]$ is the Fourier Transform operator, ν_x and ν_y are the spatial frequencies.

The accuracy of the two-flat-test method depends on the translation along each axis. The translations lead to filtering of the measured spatial frequency region. The Modulation Transfer Function of filtration for different translations is shown in Figure 3. It can be seen from the figure that the larger translations allow the reconstruction only of low frequencies because of this filtering. At the same time, smaller translations provide us with a large bandwidth of the spatial frequencies where noise limits the reconstruction.

3. Results

We have applied the two-flat-test method in numerical simulations. Initially, the accuracy of the method was estimated using the surface topography of a test flat modeled as a two dimensional sine function called harmonic and stored as a 1000×1000 matrix. During the simulation, harmonics from 1 to 240 order were retrieved. This range of the harmonics corresponds to the spatial frequency range from 1.67×10^{-3} to 4×10^{-1} mm^{-1}. Two model apertures were adopted in these initial simulations: a 600mm diameter aperture for a spatial frequency range of $1.67\times10^{-3} - 3.0\times10^{-2}$ mm^{-1} (denoted Region I); and a 33mm diameter aperture for a spatial frequency range of $3.0\times10^{-2} - 4\times10^{-1}$ mm^{-1} (denoted Region II). We defined these two ranges of spatial frequency based upon the technical characteristics. The upper bound of the first range was defined by the aperture of the flat under the measurement and the lower bound of the first range was defined by the size of the polishing tool. The lower bound of the second region was defined by the Nyquist frequency.

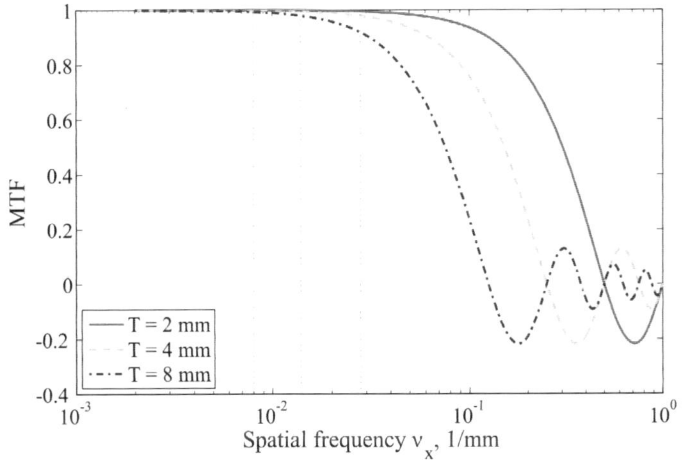

Figure 3. Modulation Transfer Function for different translation T values.

Each modeled surface topography was translated by T in both x and y directions. Having four surface height distributions (see Equation (1)), the two-flat-test method was applied. Finally, the reconstruction error was computed as follows:

$$E = \frac{\mathrm{RMS}_{test} - \mathrm{RMS}_{retrieved\_test}}{\mathrm{RMS}_{test}}, \quad (3)$$

where RMS_{test} is a root mean square value of the surface height distribution of the test flat, and $\mathrm{RMS}_{retrieved\_test}$ is a root mean square value of the surface height distribution of the retrieved test flat.

We are looking for a translation T value which guarantees a reconstruction error of less than 5%, over the defined spatial frequency regions. The results of the simulations, including the spatial frequency regions, translation T values and computed reconstruction errors are presented in Table 1.

Table 1. The result of the numerical simulations.

Spatial frequency range	Harmonic band	Translation, T (mm)	Error (%)
Region I	1-18	48 - 4,8	4.95
Region II	18-87	4.8 – 1.2	5
Region II	240	1.2	33.88

The results of the first simulation showed the theoretical limitations of the spatial frequency range of the two-flat test method.As can be seen from Table 1 the Region I wasn't totally retrieved.

Further simulations of the two-flat-test method were applied to actual surface data. Surface topography was obtained by mesurements of a test flat with aperture 100mm in diameter and a surface accuracy about λ/40 using an Intelliun H2000 interferometer [32]. Measurements were obtained with reduced sensitivity to vibration. The measured surface topography of the test flat is presented in Figure 4. The measured topography was band-pass filtered in the spatial frequency range from 1.67×10^{-3} to 3.0×10^{-2} mm^{-1} (see Figure 5).

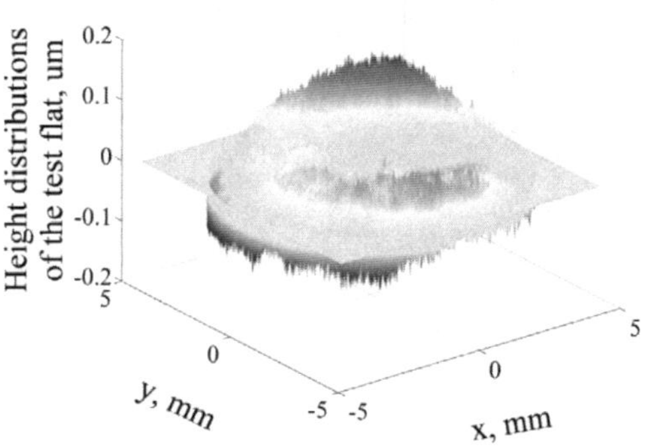

Figure 4. Initial surface topography.

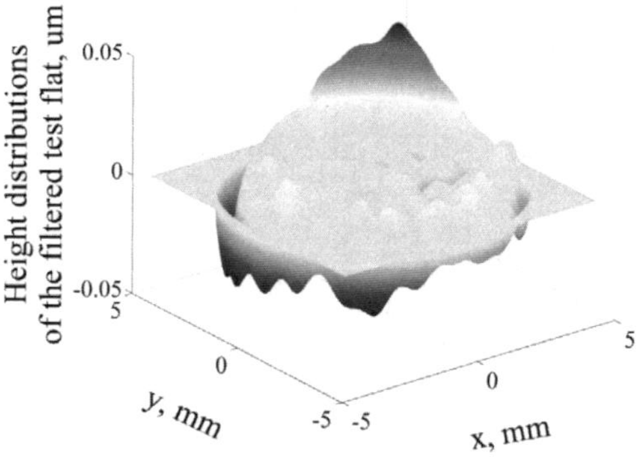

Figure 5. Initial surface topography with filtration over $1.67\text{x}10^{-3} – 3.0\text{x}10^{-2}$ mm^{-1} region.

It can be seen from the results presented in Fgures 4 and 5 that the filtered topography has been sucsessfully reconstructed by the two-flat-test method. The estimation of the reconstruction accuracy was computed using (Equation (3)). The reconstruction error and the translation T values are presented in Table 2.

Table 2. Results of simulations on actual topography data.

Spatial frequency region	Translation, T (mm)	Error (%)
Region I	0.21	5.3805
Region I	0.43	5.4184
Region I	0.87	5.507

The results of the second simulation showed the allowed translations for the two-flat-test calibration in chosen spatial frequency region with minimum reconstruction error.

4. Conclusion

In this paper we have analysed the applicability of the two-flat-test calibration method to measuring surface topography of the optical flats across a wide range of spatial frequencies.

To determine the accuracy of the method a number of numerical simulations have been perfomed. The simulation results show the translation T value which guarantees the quality of the reconstruction for the 1.67×10^{-3} – 3.0×10^{-2} mm^{-1} spatial frequency region to be 4.8mm, and to be 1.2mm for the 3.0×10^{-2} – 4×10^{-1} mm^{-1} region. Consequently, the two-flat-test method is valid for absolute calibration, in terms of desired accuracy, for spatial frequencies from 1.67×10^{-3} to 5.0×10^{-2} mm^{-1} within the simulation setup used.

The results for reconstruction of an actual surface, obtained using the experimental setup with an Intellium H2000 interferometer, show the accurate reconstruction of surface topography for a spatial frequency region 1.67×10^{-3} – 3.0×10^{-2} mm^{-1} with translation T value more than 0.2mm.

The practical importance of our study is the measurement of the surface topography and optical aberrations of large aperture optics with nanometre resolution. The results showed the possibility to apply the two-flat-test method to the measurement in a certain spatial frequency region with a great precision.

References

[1] Bray M and Liard L 1999 *Proc. SPIE* **3739** 277-786X
[2] Bouillet S 2011 *Proc. SPIE* **8169** 816914-1
[3] Campbell J H 2004 *Proc. SPIE* **5341** 84-101
[4] Born M and Wolf E 1970 *Principles of Optics. Electromagnetic Theory of Propagation, Interference and Diffraction of Light* (Oxford: Pergamon Press), 985
[5] Lord Rayleigh 1893 *Nature* **48** 212–14
[6] Chen J, Song D, Zhu R, Wang Q, and Chen L 1996 *Opt. Eng.* **35** 1936–42
[7] Powell I and Goulet E 1998 *Appl. Opt.* **37** 2579–88
[8] Vannoni M and Molesini G 2005 *Metrologia* **42** 389–93
[9] Schulz G and Schwider J 1967 *Appl. Opt.* **6** 1077–84
[10] Schulz G and Schwider J 1971 *Appl. Opt.* **10** 929–934
[11] Schwider J, Burow R, Elssner K E, Spolaczyk R, and Grzanna J 1985 *Appl. Opt.* **24** 3059-61
[12] Dew G 1966 *J. Sci. Instrum.* **43** 409-15
[13] Fritz B S 1984 *Opt. Eng.* **23** 379–83
[14] Evans C J and Kestner R N 1996 *Appl. Opt.* **35** 1015–21
[15] Dongqi S, Erlong M, Yongxin S, and Huaijiang Y 2012 *Opt. Lett.* **37** 3198-200
[16] Freischlad K 2001 *Appl. Opt.* **40** 1637–48
[17] Mantravadi M and Malacara D 2007 *Optical Shop Testing Chapter 1. Newton, Fizeau, and Haidinger Interferometers* (Hoboken: A John Wiley & Sons) chapter 1 pp 1-40
[18] Ai C and Wyant J C 1992 *Proc SPIE* **1776** 73-83
[19] Ai C and Wyant J C 1993 Appl. Opt. 32 4698–705
[20] Ai C, Wyant J C, Shao L Z and Parks R E 1996 *Method and apparatus for absolute measurement of entire surfaces of flats* U.S. patent 5,502,566
[21] Xu C, Chen L, and Yin J 2009 *Appl. Opt.* **48** 2536-41
[22] Xin J and Tingwen X 2010 *Proc. SPIE* **7656** 76563E-1
[23] Parks R 1978 *Proc. SPIE* **153** 56-63
[24] Küchel M F 2001 *Optik (Jena)* **112** 381–91
[25] Griesmann U 2006 *Appl. Opt.* **45** 5856–65
[26] Lehmann P H, Osten W and Albertazzi A 2013 *Proc. SPIE* **8788** 87882L-1
[27] Griesmann U 2006 *Appl. Opt.* **45** 5856-65
[28] Parks R E, Shao L Z and Evans C J 1998 *Appl. Opt.* **37** 5951-56
[29] Vannoni M and Molesini G 2008 *Appl. Opt.* **47** 2133–45
[30] Burke J and Griesmann U 2012 *Proc. SPIE* **8493** 84930F-1
[31] Morin F and Bouillet S 2007 *Proc. SPIE* **6616** 66164G-1
[32] Szwaykowski et al. 2012 *Interferometric system with reduced vibration sensitivity and related method* U.S. patent 2012/0026507 A1

BMSTU Seminars IOP Publishing
Journal of Physics: Conference Series **584** (2015) 012021 doi:10.1088/1742-6596/584/1/012021

Pseudo-stochastic signal characterization in wavelet-domain

Kirill I Zaytsev*, Andrei A Zhirnov, Valentin I Alekhnovich, and Stanislav O Yurchenko**

Bauman Moscow State Technical University, 105005, 2nd Baumanskaya str. 5, Moscow, Russia

E-mail: `*kirzay@gmail.com, **st.yurchenko@mail.ru`

Abstract. In this paper we present the method for fast and accurate characterization of pseudo-stochastic signals, which contain a large number of similar but randomly-located fragments. This method allows estimating the statistical characteristics of pseudo-stochastic signal, and it is based on digital signal processing in wavelet-domain. Continuous wavelet transform and the criterion for wavelet scale power density are utilized. We are experimentally implementing this method for the purpose of sand granulometry, and we are estimating the statistical parameters of test sand fractions.

1. Introduction

Characterization of pseudo-stochastic signals, containing many similar but randomly-located fragments, is highly important problem of applied physics and signal processing. Images of granulated media, complex surfaces, and biological objects are common examples of such signals [1-13].

Development of novel techniques for pseudo-stochastic signal characterization is an important topic of applied physics. Fourier-domain signal processing does not perform well at solving this problem because of complete non-locality of Fourier transform kernels in spatial-domain. Signal processing in spatial-domain is commonly used for pseudo-stochastic signal analysis, including nonlinear filtering and morphological processing [14, 15]. Spatial-domain processing performs well at solving this problem; however, it exhibits relatively high computational complexity. Wavelet-domain processing of pseudo-stochastic signals is a perspective instrument for solving the specified problem because of high localization of wavelet transform kernels both in spatial-domain and frequency-domain [16,17].

This paper presents the wavelet-domain approach for characterizing the signals, consisting of a large number of randomly located fragments. We estimate the statistical characteristics for pseudo-stochastic signals using one-dimensional (1D) continuous wavelet transform and the criterion of wavelet-scale power density. We implement the proposed technique for the purpose of sand granulometry.

Content from this work may be used under the terms of the Creative Commons Attribution 3.0 licence. Any further distribution of this work must maintain attribution to the author(s) and the title of the work, journal citation and DOI.
Published under licence by IOP Publishing Ltd

2. Estimation of pseudo-stochastic signal parameters

1D direct continuous wavelet transform of $s(x)$ is defined with the projection of function on the wavelet transform basis

$$C(a,b) = \mathcal{W}[s(x)] = \langle s(x)\psi(a,b,x)\rangle = \int_{-\infty}^{+\infty} s(x)\psi(a,b,x)dx, \qquad (1)$$
$$\mathcal{W}: R \to R^2, \qquad a,b \in R, \qquad a \neq 0$$

where $C(a,b)$ are coefficients of wavelet decomposition, and $\psi(a,b,x)$ are wavelet decomposition kernels with a-scale and b-translation. All $\psi(a,b,x)$-kernels form a basis for signal projection in $\mathcal{L}^2(R)$ wavelet-domain space. Kernels, $\psi(a,b,x)$, are constructed by scaling and shifting of the mother wavelet, $\psi(x)$:

$$\psi(a,b,x) = |a|^{-1/2}\psi\left(\frac{x-b}{a}\right). \qquad (2)$$

Figure 1 shows schematic representation of proposed algorithm for characterizing the pseudo-stochastic signals. We are finding the statistical characteristics of randomly located similar fragments of signal, $s(t)$, via post-processing of direct continuous wavelet decomposition (1). We introduce the criterion of wavelet-scale power density, $P(a)$, as an intermediate step of statistical characteristic calculation

$$P(a) = \frac{1}{a^2}\int_{-\infty}^{+\infty} |C(a,b)|^2 db. \qquad (3)$$

When $s(x)$ contains a significant number of similar but randomly located fragments with the average size $\bar{d}$, the $P(a)$-function exhibits a sharp peak centered at a_{max} with the full-width at half-maximum Δa. If the signal contains several fractions of similar objects, randomized in space, several peaks appear in $P(a)$. These peaks overlap in case of large standard deviation, σ_d, or close average sizes of fraction elements, $\bar{d}$.

In this paper, we are considering the problem of single fraction characterization, and we are utilizing $P(a)$ for calculating the statistical characteristics of single fraction: the average

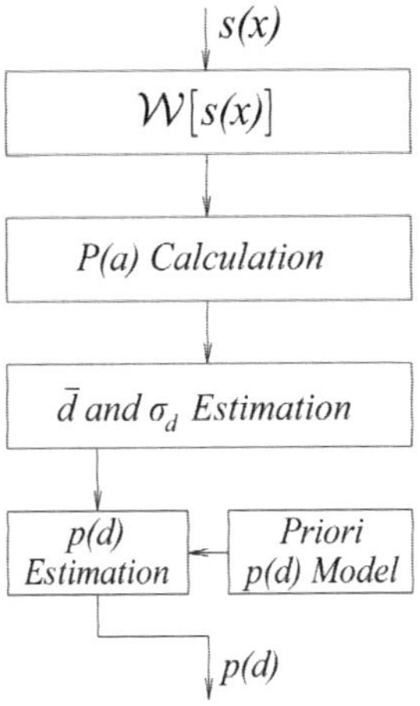

Figure 1. Schematic representation of wavelet-domain technique for characterization of pseudo-stochastic signals: $s(x)$ and $\mathcal{W}[s(x)]$ represent the input signal and its wavelet transform, respectively; $P(a)$ is wavelet-scale power density, $\bar{d}$ and σ_d are parameters of statistical distribution function $p(d)$.

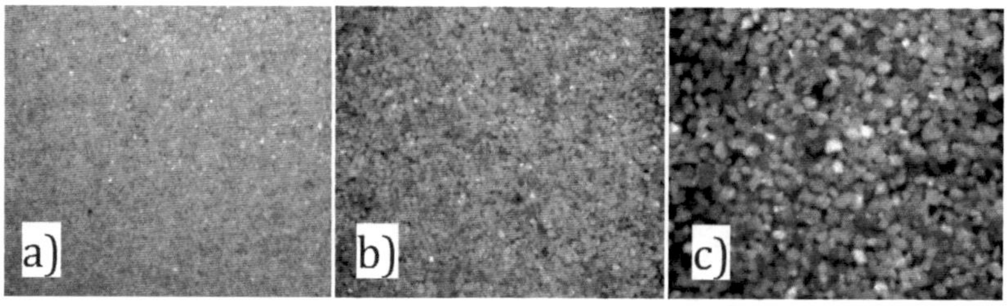

Figure 2. Digital images of sand with various average diameter of particle, $\overline{d}$: **(a)** corresponds to $\overline{d} \cong 0.45$ mm, **(b)** corresponds to $\overline{d} \cong 0.94$ mm, and **(c)** corresponds to $\overline{d} \cong 1.88$ mm.

fragment size, $\overline{d}$, and the standard deviation of fragment size, σ_d. These parameters depend on the parameters of peak in $P(a)$-function:

$$\overline{d} = f_1(a_{max}), \qquad \sigma_d = f_2(\Delta a), \tag{4}$$

where $f_1(...)$, $f_2(...)$ are some calibration functions.

3. Algorithm implementation for sand granulometry

We are implementing the proposed technique for studying the set of test sand fractions having calibrated grain diameters. Table 1 shows the sand fraction parameters ($\sigma_d \cong 1/6(d_{max} - d_{min})$), and figure 2 shows digital images of these sand fractions.

We use the 1D wavelet decomposition of certain image row, $s(x) \equiv s(x, y = y')$, instead of 2D wavelet decomposition of entire image, $s(x, y)$, in order to reduce the computational complexity. Each row represents the cross-section of the object textures, $y = y'$, and contains complete information about the inhomogeneities of pseudo-stochastic object. Moreover, 1D approximation is much faster than the 2D approach, allowing us real-time data processing. Figure 3 shows the result of direct continuous wavelet transform (1) of image rows. The difference between wavelet spectra can be observed directly from the result of the wavelet decomposition, but this it is difficult to estimate this difference without data post-processing.

We calculate the 1D criterion of wavelet-scale power density (3) for all the wavelet spectrums (figure 3) of sand samples, and figure 4 shows the results. The curves of figure 4 exhibit sharp maximum centered at a_{max} and having the full-width at half-maximum Δa. We are estimating a_{max} and Δa for each peak; Table 2 shows the results. On the step of system calibration (we study various test samples of *priory* characterized sand fractions for this purpose) we estimate the following polynomial approximation of calibration functions, $\overline{d} = f_1(a_{max})$ and $\sigma_d = f_2(\Delta a)$:

$$\begin{aligned} \overline{d} &= -8.104 + 0.916a_{max} - 0.021a_{max}^2, \\ \sigma_d &= 0.247 - 0.041\Delta a + 0.018\Delta a^2. \end{aligned} \tag{5}$$

Table 1. *Priory* parameters of sand fractions.

#	$\overline{d}$, mm	$d_{min}...d_{max}$, mm
1	0.45	0.35...0.55
2	0.94	0.63...1.25
3	1.88	1.26...2.50

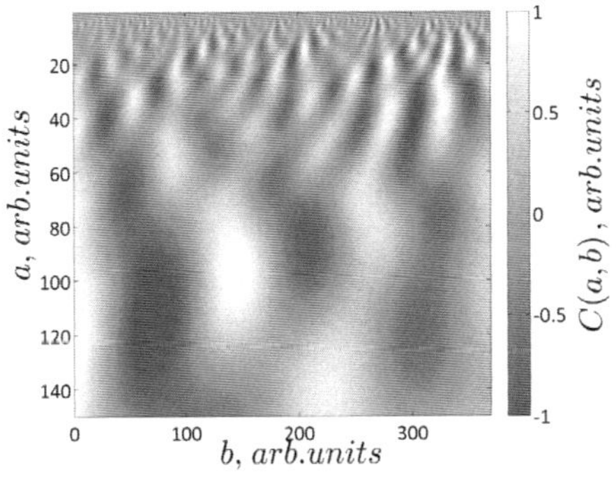

a)

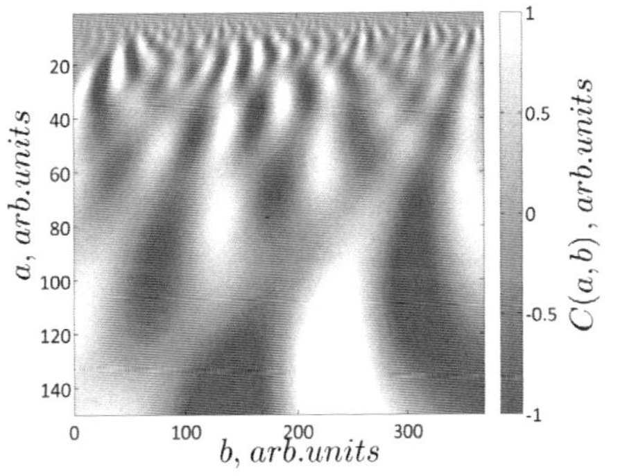

b)

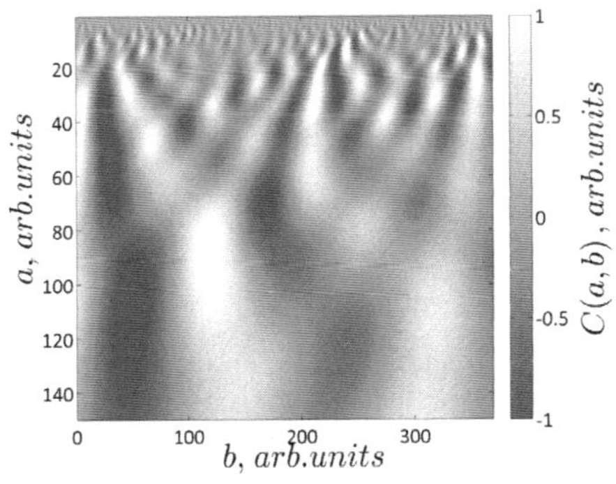

c)

Figure 3. 1D continuous wavelet transform of appropriate rows of digital images (figure 2): **(a)** corresponds to $\overline{d} \cong 0.45$ mm sand fraction, **(b)** corresponds to $\overline{d} \cong 0.94$ mm sand fraction, and **(c)** corresponds to $\overline{d} \cong 1.88$ mm sand fraction. Daubechies mother wavelet $'db8'$ is applied for wavelet decomposition.

Nonlinearity of (5) occurs due to the large number of factors, including aberrations of optical system, strongly correlated noises, overlapping of separate grains in digital images, and decay of optical modulation transfer function with an increase of spatial frequency of image textures.

Table 2 shows results of estimation of sand characteristics, $\overline{d}'$ and σ'_d, using equation (5). We

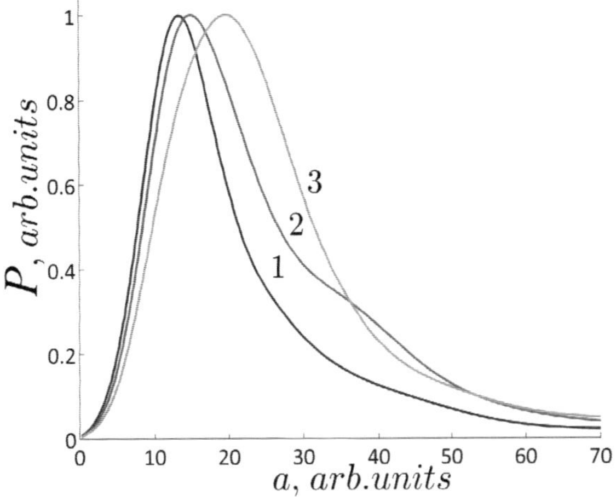

Figure 4. Wavelet-scale power density criterion, $P(a)$, calculated for various sand fractions: (1) corresponds to $\overline{d} \cong 0.45$ mm sand fraction, (2) corresponds to $\overline{d} \cong 0.94$ mm sand fraction, and (3) corresponds to $\overline{d} \cong 1.88$ mm sand fraction.

Table 2. Estimated parameters of sand fractions.

#	$\overline{d}$, mm	a_{max}, arb. units	$\overline{d}'$, mm	$\delta_{\overline{d}}$
1	0.450	13.25	0.430	0.05
2	0.940	15.00	0.946	0.01
3	1.880	19.75	1.979	0.01
			$\langle\delta_{\overline{d}}\rangle = 0.04$	

#	σ_d, mm	Δa, arb. units	σ_d', mm	δ_σ
1	0.033	13.50	0.028	0.15
2	0.103	18.25	0.105	0.02
3	0.207	21.59	0.206	0.01
			$\langle\delta_\sigma\rangle = 0.11$	

are also estimating the relative errors of statistic parameter determination

$$\delta_{\overline{d}i} = (\overline{d}_i' - \overline{d}_i)/\overline{d}_i, \qquad \delta_{\sigma i} = (\sigma_{di}' - \sigma_{di})/\sigma_{di},$$
$$\langle\delta_{\overline{d}}\rangle = \sqrt{\frac{1}{N}\sum_{i=1}^{N}\delta_{\overline{d}i}^2}, \qquad \langle\delta_\sigma\rangle = \sqrt{\frac{1}{N}\sum_{i=1}^{N}\delta_{\sigma i}^2}, \tag{6}$$

where i is the sample index, and $N = 3$ stands for the total number of samples. The accuracy of the statistical parameter estimation is much higher for 2nd and 3rd samples compared to the 1st one due to the limitations of the imaging system resolution. The higher the contrast in sand images we observe, the higher the accuracy of statistics reconstruction we achieve.

We could implement real-time signal processing with a repetition rate of algorithm implementation higher than 50 Hz since 1D continuous wavelet transform is applied. The repetition rate can be significantly raised by increasing the camera frame rate and via the parallel processing of pseudo-stochastic data. This would allow us implementation of the method in real-time technological applications, such as remote sensing in a fast flow of microparticles.

4. Conclusion

We have proposed fast and precise method for characterizing the pseudo-stochastic signals based on digital signal processing in wavelet-domain. We have implemented this technique for processing the digital images of various sand fractions as a representative example.

Acknowledgments

This work is supported by the Russian Ministry of Education and Science, Project # 3.1526.2014/K.

References

[1] Bertrand D, Robert P, Melcion J P, and Sire A 1991 *Powder Technology* **66** 171–6
[2] Sheng J, Malkiel E, and Katz J 2006 *Appl. Opt.* **45** 3893–901
[3] Staszewski W J, Pierce S G, Worden K, Philp W R, Tomlinson G R, and Culshaw B, 1997 *Opt. Eng.* **36** 1877–88

[4] Zaytsev K I, Katyba G M, Yakovlev E V, Gorelik V S, and Yurchenko S O 2014 *J. Appl. Phys.* **115** 213505
[5] Wang Z and Ma H 2006 *Opt. Eng.* **45** 045601
[6] ZadoroznyA, Zhang H, and Jagersand M 2002 *Proc. 15th International Conference on Vision Interface* s8
[7] Yeo J-B, Yun S-D, and Lee H-Y 2007 *J. Appl. Phys.* **102** 084502
[8] Guo Q, Chen , Smith B C, and Jhon M S 2007 *J. Appl. Phys.* **97** 10P301
[9] Peled A, Baranauskas V, Rodrigues C, ArtWeisman D, Grantman L, and Friesem A A 1995 *J. Appl. Phys.* **77** 6208
[10] Zaytsev K I, and Yurchenko S O 2014 *Appl. Phys. Lett.* **105** 051902
[11] Maire E, Colombo P, Adrien J, Babout L, and Biasetto L 2007 *J. European Ceramic Society* **27** 1973
[12] Brosnan T and Sun D-W 2004 *J. Food Engineering* **61** 3
[13] Morfill G E and Ivlev A V 2009 *Rev. Mod. Phys.* **81** 1353–404
[14] Chen Y and Dougherty E R 1994 *Opt. Eng.* **33** 2713–22
[15] Dougherty E R, Newell J T, and Pelz J B 1992 *Pattern Recognition* **25** 1181
[16] I Daubechies 1990 *IEEE Transact. Information Theory* **36** 961-1005
[17] Zaytsev K I, Zhirnov A A, and Yurchenko S O *Opt. Eng.* (Under Review, 2014)

Difference Raman spectroscopy of DNA molecules

Andrey S Anokhin[1], **Vladimir S Gorelik**[2,3], **Galina I Dovbeshko**[4], **Alexander Yu Pyatyshev**[3] **and Yury I Yuzyuk**[1]

[1] Southern Federal University, Rostov-on-Don, Russia

[2] P.N. Lebedev Physical Institute of the Russian Academy of Sciences, Moscow, Russia

[3] Bauman Moscow State Technical University, Moscow, Russia

[4] Institute of physics National Academy of Sciences of Ukraine, Kyiv, Ukraine

E-mail: gorelik@sci.lebedev.ru

Abstract. In this paper the micro-Raman spectra of calf DNA for different points of DNA sample have been recorded. The Raman spectra were made with help of difference Raman spectroscopy technique. Raman spectra were recorded with high spatial resolution from different points of the wet and dry samples in different spectral range (100÷4000cm^{-1}) using two lasers: argon (514.5 nm) and helium -neon (632.8 nm). The significant differences in the Raman spectra for dry and wet DNA and for different points of DNA molecules were observed. The obtained data on difference Raman scattering spectra of DNA molecules may be used for identification of DNA types and for analysis of genetic information associated with the molecular structure of this molecule.

1. Introduction and background

Important components of biological objects are nucleotides, which enter the composition of deoxyribonucleic acid (DNA) and ribonucleic acid (RNA). The fundamental function of these molecules is to keep and transfer the information that is necessary for existence and reproduction of biological objects. Therefore, any modification of the nucleotide structure of an object affects strongly its vital functions. A structural modification of DNA and genes changes also the genetic information. The microstructure of biological objects can be modified by different chemical and physical impacts, in particular, under electromagnetic irradiation. Methods of molecular spectroscopy, including Raman scattering and photoluminescence, are widely used to study the structural modification of polynucleotides.

A large number of studies are devoted to DNA Raman spectra recording. Raman spectra of an aqueous solution (pH=7) of calf thymus DNA in a frequency range of 600 to 1800 cm^{-1} and at temperatures of 25, 80, and 98^{0}C, in the presence of 0.01 mol/l sodium cacodylate and 0.001 mol l^{-1} ethylenediaminetetraacetic acid were reported in [1, 2]. The total concentration of Na^+ ions was 0.0075 mol l^{-1}. It was found that the intensities of individual lines due to internal vibrations of all DNA bases increased when approaching the melting temperature (80^{0}C). The Raman spectra of both an aqueous solution of calf thymus DNA and dry DNA were investigated in the frequency range of 600 to 1800 cm^{-1} at temperatures of -100, 20, and 130^{0}C [3]. The frequencies corresponding to adenine and guanine were found to redshift upon cooling, while the frequencies corresponding to thymine and cytosine remained invariable.

Content from this work may be used under the terms of the Creative Commons Attribution 3.0 licence. Any further distribution of this work must maintain attribution to the author(s) and the title of the work, journal citation and DOI.
Published under licence by IOP Publishing Ltd

The conformational changes in DNA aqueous solutions with Na^+, Cs^+, Mn^{2+}, and Mg^{2+} counter ions introduced into different sites in the macromolecule were studied in [4–6]. These counterions neutralize phosphate groups in the double helical backbone of the DNA molecule. As a result, low frequency Raman spectra in the range of 60 to 140 cm^{-1} were also recorded.

In [1] and [7], the data on changes in the Raman spectra at conformational transitions of DNA molecules in aqueous solutions by the addition of NaCl salt at a concentration of 3-4% and a relative humidity of 75 and 92% in the frequency range 600-1800 cm^{-1} were presented. It was found that when the relative humidity exceeds 92% of the DNA transferred from A- to B-form.

In [8-10] the influence of aspirin, Adriamycin and ethidium bromide on DNA Raman spectrum has been investigated. It was studied the changes DNA structure changing as a result of double helix interaction with these compounds. Accomplished studies are important in terms of influence rhenium aspirin microstructure DNA, as well as in view of the fact that Adriamycin used for cancer treatment, and ethidium bromide is a potent mutagen. It became clear from these studies, that in the frequency range of 600-1800 cm^{-1} substantial change have undergone, which indicates on significant interaction of DNA molecules with these compounds.

Despite a large number of studies devoted to DNA Raman spectra, the low-frequency (0 to 600 cm^{-1}) and high-frequency (to 4000 cm^{-1}) spectral ranges have been poorly studied. Completed to present days studies of the Raman spectra of DNA were performed in macroscopic volumes of test samples. The purpose of our paper was to analyzing of DNA Raman spectra in a wide frequency range (100 to 4000 cm^{-1}) with high spectral resolution (1 cm^{-1}) for different points of the sample. In this work we used calf DNA samples and have studied the effect of water influence on the Raman spectrum.

DNA is a biopolymer comprised of nucleotides as monomers [11 - 13]. Each nucleotide consists of a phosphoric acid residue attached to a deoxyribose sugar to which is attached through a glycosidic bond one of the four nitrogenous bases adenine (A), guanine (G), cytosine (C) and thymine (T). DNA polymer has a fairly complex structure [14-17]. The skeleton of each of the DNA strands is composed of alternating sugar and phosphate [16].

2. Results

Raman spectra were excited and recorded with using of typical experimental micro-Raman setup. Laser beam was directed by a transparent mirror onto the substrate with the material studied. The secondary radiation (Raman signal) was directed (using a semitransparent mirror) to the input slit of a spectrograph Renishaw InVia Reflex. Spectrograph equipped with a microscope Leica DMLM, having a resolution of ~1 μm and lens 5x, 20x, 50x and 100x. An argon and helium-neon lasers with wavelength of 514.5 and 632.8 nm generations were used as an excitation source.

Figure 1 shows the Raman spectrum obtained from three different points of dry DNA when excited by an argon laser. From this figure we can see that in the Raman spectra present background photoluminescence. This pattern is different for different points of the target DNA. We can see the essential Raman spectra difference from each other at high frequencies. Near the edge of the low frequency (~ 100 cm^{-1}) there is the peak, which was investigated at room and liquid nitrogen temperatures before [18, 19].

Figure 2 shows Raman spectra of wet DNA obtained from three different points with excitation by helium-neon laser. As we can see the Raman spectra is on photoluminescence background. The interaction of DNA with water results in the some changes of several bands frequencies, corresponding to PO_2^-, C-O and C-C vibrations, as well as all components of DNA nucleic bases [3, 20].

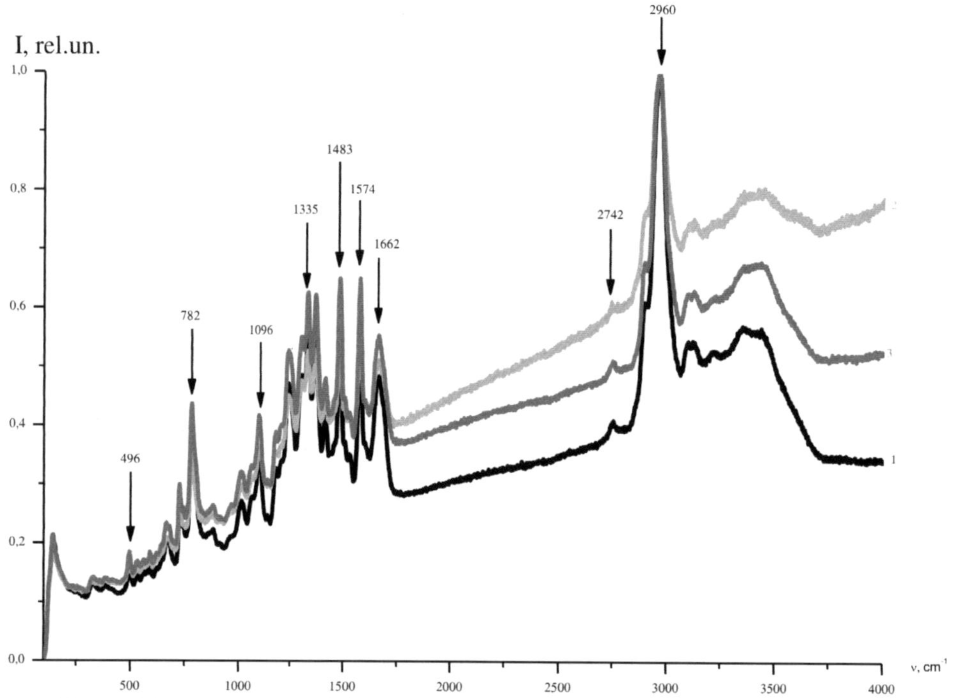

Figure 1. Raman spectra of dry DNA excited by an argon laser: 1, 2, and 3 corresponds to first, second, and third points of DNA molecule, respectively.

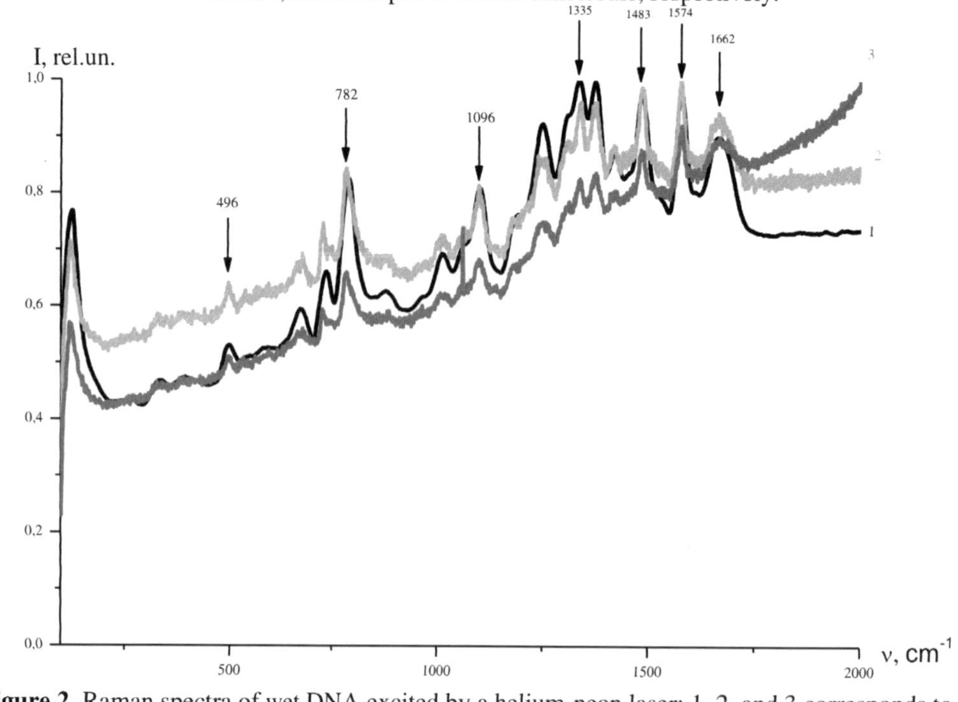

Figure 2. Raman spectra of wet DNA excited by a helium-neon laser: 1, 2, and 3 corresponds to first, second, and third points of DNA molecule, respectively.

To establish quantitative differences in the Raman spectra obtained from different points of the DNA dry and wet samples we were constructed difference spectra [11, 21-24] using the ratio of

$$K_X^B(\nu)=1-\left|I_X(\nu)-I_B(\nu)\right|. \quad (1)$$

Here $I_X(\nu)$ and $I_B(\nu)$- the normalized Raman spectra of the current point (X) and the base point (B). For the base point adopted the first point. On the basis of the difference spectra were calculated integral coefficients of compliance with the following formula

$$K=\frac{\sum_{i=1}^{N}\left(1-|I_X(\nu)-I_B(\nu)|\right)_i}{N}. \quad (2)$$

Figure 3 shows the difference Raman spectra excited by argon laser between the second and first and third and first points of the dry DNA. Figure 4 illustrate the difference Raman spectra of wet DNA obtained for the two points (the second and first and third and first). In this case the excitation of Raman spectra was with using of helium-neon laser. The frequency range is indicated by arrows.

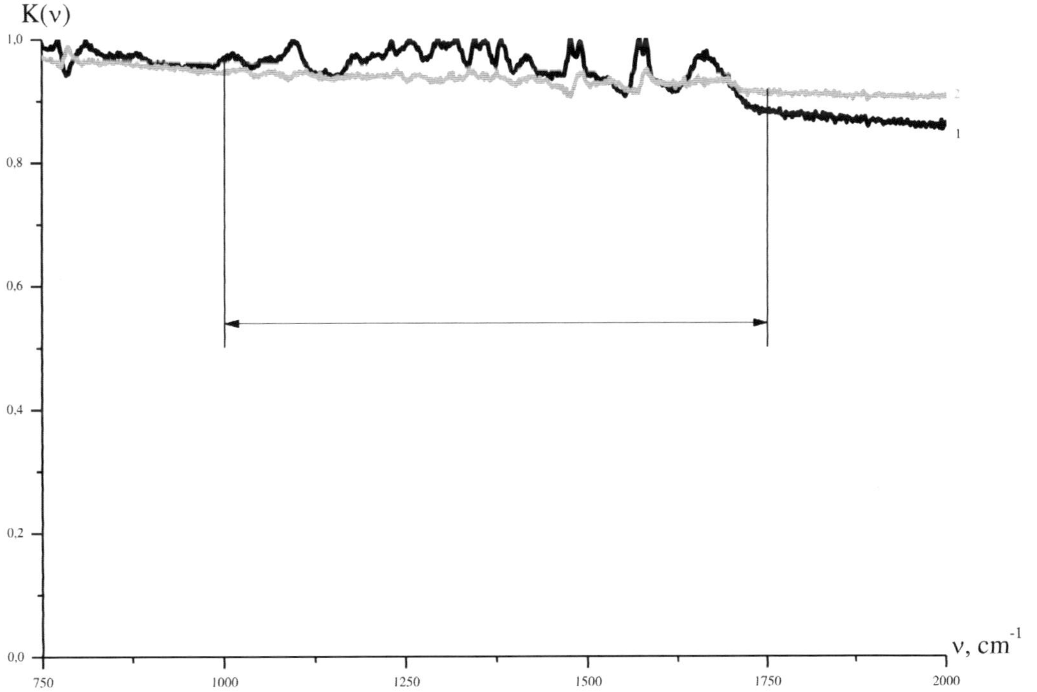

Figure 3. Difference Raman spectra of dry DNA when excited by an argon laser: 1 corresponds to second and first points, with a coefficient of compliance of K = 0.91; 2 corresponds to third and first points, with a coefficient of compliance of K = 0.95.

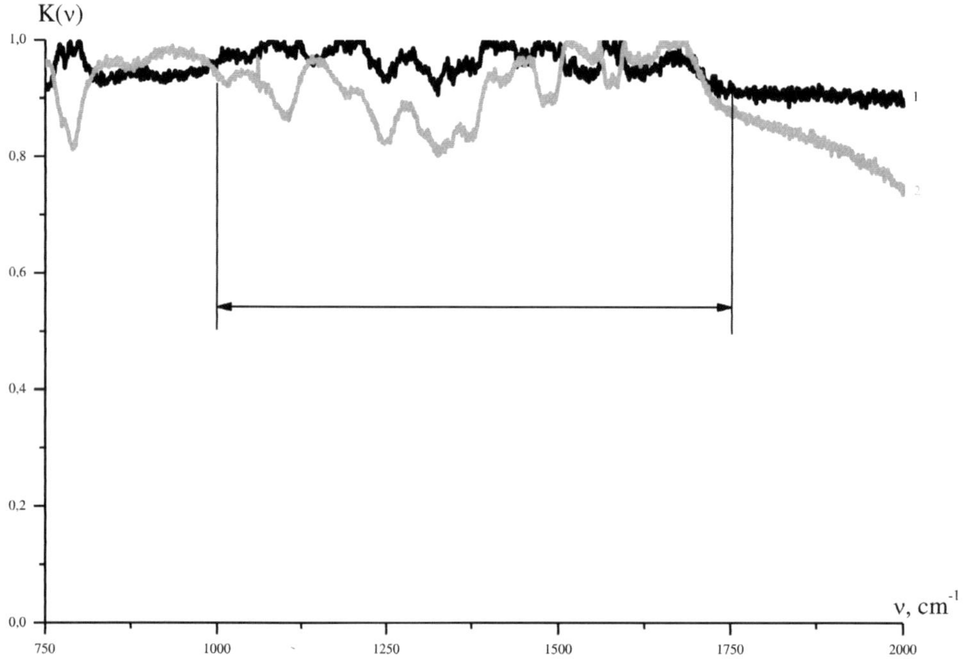

Figure 4. Difference spectra of wet DNA when excited by a helium-neon laser: 1corrsponds to first and second points, with a coefficients of compliance of K = 0.96; 2 corresponds to third and first points; with a coefficients of compliance of K = 0.89.

As can be seen from figure 4, Raman spectra of wet DNA at different points differ considerably, indicating that the DNA strongly interacts with water.

3. Conclusions

Thus, in this paper Raman spectra of dry and wet calf DNA in a wide frequency range with high spatial resolution (~10 μm) have been examined. By using difference spectroscopy technique we have revealed the significant difference in theses spectra, recorded from different points of dry and wet DNA. Obtained results on difference Raman spectroscopy of DNA molecules is interesting for DNA characterizing and also for the study of DNA structural changing as a result of different type perturbations.

Acknowledgments

This work was supported by RFBR (grants 12-02-00491, 13-02-00449, 13-02-90420, 14-02-00190, 14-02-90406, 14-02-90007).

References

[1] Carey P R 1982 *Biochemical applications of Raman and Resonanse Raman spectroscopies* (New York: Academic Press)
[2] Erfurth S C, Peticolas W L 1975 *Biopolymers.* **14**(2) 247-64
[3] Dong R, Yan X, Pang X, Liu S 2004 *Spectrochim Acta A Mol Biomol Spectrosc.* **60**(3) 557-61
[4] Muntean C M, Misselwitz R and Welfle H 2006 *Spectroscopy* **20**(5-6) 261-68
[5] Perepelytsya S M, Volkov S N 2010 *Ukr. J. Phys.* **55**(11) 1182-88
[6] Perepelytsya S M, Volkov S N 2011 *J. Molecular Liquids* **164**(1-2) 113-19

[7] Erfurth S C, Bond P J, Peticolas W L 1975 *Biopolymers.* **14**(6) 1245-57
[8] Angeloni L, Smulevich G and Marzocchi M P 1982 *Spectrochim. Acta* **38A**(2) 213-17
[9] Neault J F, Naoui M, Manfait M, Tajmir-Riah H A 1996 *FEBS Lett.* **382**(1-2) 26-30
[10] Blyzniuk Iu N, Bolbukh T V, Kruglova O B, Semenov M A, Maleev V Ya 2009 *Biopolym. Cell.* **25**(2) 126-32
[11] Duguid J G, Bloomfield V A, Benevides J M and Thomas G J Jr 1996 *Biophys J.* **71**(6) 3350-60
[12] Alberts B, Johnson A, Lewis J, Raff M, Roberts K and Walter P 2002 *Molecular Biology of the Cell* (New York: Garland Science)
[13] Butler J M 2001 *Forensic DNA Typing: Biology and Technology behind STR Marker* (London: Academic Press)
[14] Watson J D, Crick F H C 1953 *Nature* **171**(4356) 737-38
[15] Berg J M, Tymoczko J L and Stryer L 2012 *Biochemistry* (New York: W. H Freeman and Company)
[16] Ghosh A, Bansal M 2003 *Acta Crystallogr D Biol Crystallogr.* **59**(4) 620-26
[17] Lamba Om P, Wang A H- J and Thomas G J Jr 1989 *Biopolymers.* **28**(2) 667-78
[18] Gorelik V S, Dovbeshko G I, Krylov A S, Pyatyshev A Yu 2014 *Herald of Bauman Moscow State Technical University. Series: "Natural Sciences"* **1** 27-34
[19] Gorelik V S, Pyatyshev A Yu 2014 *Phys. of Wave Phenomena* **22**(4) 1-3
[20] Olsztynska-Janus S, Szymbrorska-Malek K, Gasior-Glogowska M, Walski T, Komorowska M, Witkiewicz W, Pezowicz C, Kobielarz M, Szotek S 2012 *Acta of Bioeng Biomech.* **14**(4) 121-33
[21] Langlais M, Tajmir – Riahi H A and Savoie R 1990 *Biopolymers.* **30**(7-8) 743-52
[22] Kint S, Tomimatsu Y 1979 A *Biopolymers.* **18**(5) 1073-79
[23] Benevides J M, Overman S A and Thomas G J Jr 2005 *J. Raman Spectrosc.* **36**(4) 279-99
[24] Gorelik V S, Pyatyshev A Y 2012 *Herald of Bauman Moscow State Technical University. Series: "Natural Sciences"* **2** 12-17

Journal of Physics: Conference Series **584** (2015) 012023 doi:10.1088/1742-6596/584/1/012023

In vivo spectroscopy of healthy skin and pathology in terahertz frequency range

Kirill I Zaytsev[1], Konstantin G Kudrin[2], Igor V Reshetov[3], Arseniy A Gavdush[1], Nikita V Chernomyrdin[1], Valeriy E Karasik[1] and Stanislav O Yurchenko[1]

[1]Bauman Moscow State Technical University, 105005, 2nd Baumanskaya str. 5, Moscow, Russia;
[2]P.A.Hertzen Moscow Research Institute of Oncology, 125284, 2nd Botkinskiy proezd 3, Moscow, Russia;
[3]Institute of Improvement of Professional Skill of the Federal Medico-Biological Agency of Russia, 125371, Volokolamskoye shosse 91, Moscow, Russia

E-mail: kirzay@gmail.com

Abstract. Biomedical applications of terahertz (THz) technology and, in particular, THz pulsed spectroscopy have attracted considerable interest in the scientific community. A lot of papers have been dedicated to studying the ability for human disease diagnosis, including the diagnosis of human skin cancers. In this paper we have studied the THz material parameters and THz dielectric properties of human skin and pathology *in vivo*, and THz pulsed spectroscopy has been utilized for this purpose. We have found a contrast between material parameters of basal cell carcinoma and healthy skin, and we have also compared the THz material parameters of dysplastic and non-dysplastic pigmentary nevi in order to study the ability for early melanoma diagnosis. Significant differences between the THz material parameters of healthy skin and pathology have been detected, thus, THz pulsed spectroscopy promises to be become an effective tool for non-invasive diagnosis of skin neoplasms.

1. Introduction

The problem of skin cancer diagnosis is of high importance in medical science, and novel methods based on modern physical principles are being developed in order to solve this problem. Recently terahertz (THz) technology, in particular, THz pulsed spectroscopy [1,2], attracts considerable interest of scientific community as a possible instrument of medical diagnosis [3–36].

THz pulsed spectroscopy appeared at 1975 as a result of Auston's research of semiconductor photoconductivity under the femtosecond optical pumping [2]. It was developing at the end of the XX century with an appearance of highly-stable femtosecond lasers, and many techniques for THz pulse generation and detection appeared [1]. The methods for THz material parameter characterization [3–11] and algorithms for THz time-of-flight tomographic studying the internal structure of sample by means of THz pulsed spectroscopy [12–15] have been developed. THz pulses spectroscopy has been applied for the purposes of medical diagnosis. It has been implemented for studying the skin cancers, and the ability for diagnosing the basal cell carcinoma have been demonstrated [16,17]. Numerous researches have been dedicated to diagnosis of colon tissue cancer [18,19], to intraoperative diagnosis of breast cancer [20,21], to sensing in

Content from this work may be used under the terms of the Creative Commons Attribution 3.0 licence. Any further distribution of this work must maintain attribution to the author(s) and the title of the work, journal citation and DOI.
Published under licence by IOP Publishing Ltd

corneal tissue [22], enamel tissue [23], liver [24], and blood [25], and to studying the skin burns [26,27]. In paper [28] the intrinsic biomarkers for non-melanoma skin cancer detection with THz spectroscopy have been introduced. Possibilities for bacteria detection via THz spectroscopy [29] and for THz molecular spectroscopy [30–33] have been introduced.

Recently, a lot of attention is paid to early diagnosis of melanoma, which is reported to be the most dangerous skin cancer [34], and THz pulsed spectroscopy should be considered as a possible method for solving this problem [35]. The present paper is dedicated to studying the THz material parameters and THz dielectric properties of healthy skin and pathology *in vivo*. We are studying and analyzing the THz material parameters of basal cell carcinoma and healthy skin, as well as material parameters of dysplastic and non-dysplastic pigmentary skin nevi. Thus, the ability for early non-invasive diagnosis of skin neoplasms is considered.

2. Materials and methods

Let us describe an experimental set-up and the method for reconstructing the THz material parameters of healthy skin and pathology *in vivo*. We are utilizing compact THz pulsed spectrometer, based on generation of THz pulses in photoconductive antenna and detection of THz field in electrooptical detector [14,36]. Detail description of set-up operation principle is out of the present paper scope; it could be found in papers [14,36]. The set-up allows determination of complex amplitude reflectivity of *in vivo* tissue in range from 0.1 to 2.5 THz.

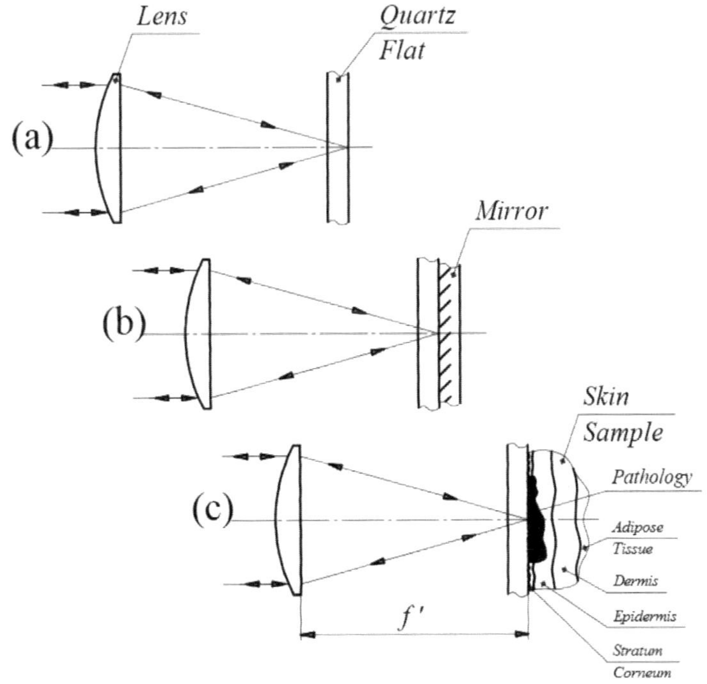

Figure 1. Schematic representation of THz waveform detection during the reflection mode sample characterization: **(a)** corresponds to measuring the empty SiO_2-flat, **(b)** corresponds to measuring the SiO_2-flat with the gold mirror placed behind it, and **(c)** corresponds to measuring the SiO_2-flat with the sample placed behind it. Note, THz beam is focused on the interface between the SiO_2-flat and the sample of interest.

The procedure for reconstructing the material parameters of *in vivo* tissue (sample) could be described as follows. Figure 1 schematically represents the process of measurements, where thick quartz flat is utilized for sample fixation. Three THz waveforms are being detected: **(a)** corresponds to the reference THz waveform reflected by the empty SiO_2-flat, $E_r(t)$; **(b)** corresponds to the THz waveform reflected by the flat with the gold mirror placed on its backside, $E_m(t)$; and **(c)** corresponds to the sample THz waveform, reflected by the flat with the sample of interest placed behind it, $E_s(t)$. Let us define Fourier spectrums of $E_r(t)$, $E_m(t)$, and $E_s(t)$ with $\widetilde{E}_r(\nu)$, $\widetilde{E}_m(\nu)$, and $\widetilde{E}_s(\nu)$.

Reconstruction of sample material parameters, $\widetilde{n}_2(\nu)$, could be performed via the minimization of an error functional

$$\widetilde{n}_2(\nu) = \arg\min_{\widetilde{n}_2(\nu)} \left[\Phi\left(\nu, \widetilde{n}_2(\nu)\right)\right] = \arg\min_{\widetilde{n}_2(\nu)} \left[M\left(\nu, \widetilde{n}_2(\nu)\right) + A\left(\nu, \widetilde{n}_2(\nu)\right)\right], \tag{1}$$

where ν is an electromagnetic wave frequency,

$$\begin{aligned} M\left(\nu, \widetilde{n}_2(\nu)\right) &= \left|\left|\widetilde{H}_{exp}(\nu)\right| - \left|\widetilde{H}_{th}\left(\nu, \widetilde{n}_2(\nu)\right)\right|\right|^2, \\ A\left(\nu, \widetilde{n}_2(\nu)\right) &= \left|\phi\left[\widetilde{H}_{exp}(\nu)\right] - \phi\left[\widetilde{H}_{th}\left(\nu, \widetilde{n}_2(\nu)\right)\right]\right|^2, \end{aligned} \tag{2}$$

$\widetilde{H}_{exp}(\nu)$ and $\widetilde{H}_{th}(\nu, \widetilde{n}(\nu))$ are experimental and theoretical transfer functions corresponding to the reflection mode measurements, $|...|$ and $\phi[..]$ are operators, which extract modulus and phase of complex functions. Experimental transfer function has the form

$$\widetilde{H}_{exp}(\nu) = \frac{\widetilde{E}_s(\nu) - \widetilde{E}_r(\nu)}{\widetilde{E}_m(\nu) - \widetilde{E}_r(\nu)}, \tag{3}$$

and theoretical transfer function is based on the model, which assumes multiple THz wave reflections within the SiO_2-flat

$$\widetilde{H}_{th}(\nu) = \frac{\widetilde{R}_{12} - \widetilde{R}_{10} + \sum_{j=1}^{N}\left(\widetilde{R}_{12}^{j+1} - \widetilde{R}_{10}^{j+1}\right)\widetilde{P}_1^{2j}\widetilde{R}_{10}^{j}}{\widetilde{R}_{13} - \widetilde{R}_{10} + \sum_{j=1}^{N}\left(\widetilde{R}_{13}^{j+1} - \widetilde{R}_{10}^{j+1}\right)\widetilde{P}_1^{2j}\widetilde{R}_{10}^{j}}, \tag{4}$$

In equation (4) operator $\widetilde{P}_1 = \widetilde{P}_1(l, \nu)$ is based on the Beer-Lambert-Bouguer law

$$\widetilde{P}_\beta(z, \nu) = exp\left(-i\frac{2\pi\nu}{c}\widetilde{n}_\beta(\nu)\, z\right), \tag{5}$$

describing the light propagation through the β dissipative medium along certain distance, z, and c is a speed of light in vacuum. Operators $\widetilde{R}_{10} = \widetilde{R}_{10}(\nu)$, $\widetilde{R}_{12} = \widetilde{R}_{12}(\nu)$, $\widetilde{R}_{13} = \widetilde{R}_{13}(\nu)$ are based on the Fresnel formulas

$$\widetilde{T}_{\xi\zeta}(\nu) = \frac{2\widetilde{n}_\xi(\nu)}{\widetilde{n}_\xi(\nu) + \widetilde{n}_\zeta(\nu)}, \qquad \widetilde{R}_{\xi\zeta}(\nu) = \frac{\widetilde{n}_\zeta(\nu) - \widetilde{n}_\xi(\nu)}{\widetilde{n}_\xi(\nu) + \widetilde{n}_\zeta(\nu)}, \tag{6}$$

describing light transmission and reflection at the interface of ξ and ζ media ($\widetilde{n}_0(\nu) = 1.0$ is a *priori* known complex refractive index of air, $\widetilde{n}_1(\nu)$ is a *priori* known complex refractive index of SiO_2, $\widetilde{n}_2(\nu)$ is a complex refractive index of sample, and $\widetilde{n}_3(\nu) = 4 \cdot 10^3 - i10^3$ is a *priori* known complex refractive index of mirror medium). In (4) N corresponds to the number of multiple wave reflections in a SiO_2-flat, and l corresponds to *priori* known thickness of reference quartz window. Note, N is limited by the width of spectrometer time-domain window, thus, only $N = 1, 2$ satellite pulses could be detected utilizing THz pulsed spectrometer.

Described procedure allows estimating the sample material parameters, $\widetilde{n}_2(\nu)$, via the reflection mode measurements, and we are implementing it for studying the material parameters of healthy skin and pathology *in vivo*.

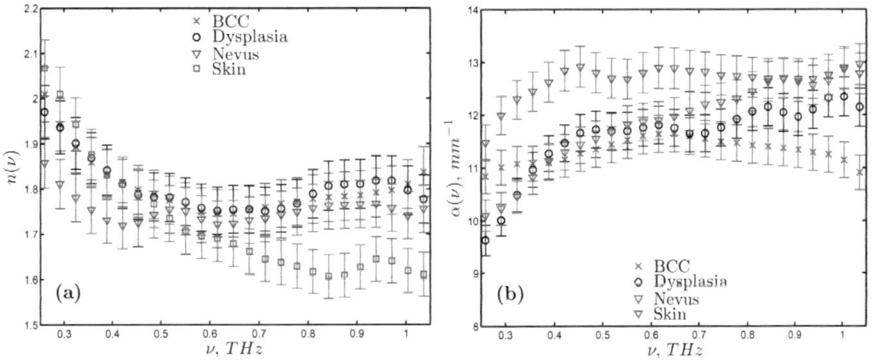

Figure 2. THz material parameters of healthy skin and pathology *in vivo*: basal cell carcinoma (×), dysplastic pigmentary nevus (○), non-dysplastic pigmentary nevus (▽), and healthy skin (□), where **(a)** corresponds to refractive index, and **(b)** corresponds to absorbtion coefficient.

3. Results

We have examined healthy skin and pathology *in vivo*. Figure 2 shows the material parameters of basal cell carcinoma, dysplastic and non-dysplastic pigmentary nevi, and healthy skin; **(a)** corresponds to the refractive index, $n_2(\nu)$, and **(b)** corresponds to the absorbtion coefficient $\alpha_2(\nu)$ ($\alpha_2(\nu)$ corresponds to light intensity absorbtion). The following equation defines the relation between $n_2(\nu)$, $\alpha_2(\nu)$, and complex refractive index of sample, $\widetilde{n}_2(\nu)$,

$$\widetilde{n}_2(\nu) = n_2(\nu) + i\frac{\alpha_2(\nu)\,c}{4\pi\nu}. \tag{7}$$

The reliable spectral range for characterizing the material parameters of *in vivo* tissue is limited both at low-frequency and high-frequency sides. The finite lateral size of skin pathology limits the spectral range at the low-frequency side due to the diffraction limit of THz beam spot. The roughness of skin surface limits the spectral range at the high-frequency side due to the strong scattering of high-frequency spectral components of THz beam on tissue surface inhomogeneities. We have experimentally found that the reliable range is located between 0.25 and 1.05 THz. Observed curves of THz material parameters (figure 2) are statistically distinguishable in low-frequency (< 0.5 THz) and high-frequency (> 0.7 THz) ranges.

Let us consider another approach for experimental data representation. Ordinary it is customary to plot the frequency dependencies of sample material parameters, $n_2(\nu)$ and $\alpha_2(\nu)$, or the frequency dependencies of sample dielectric characteristics, $\varepsilon'(\nu)$ and $\varepsilon''(\nu)$, ($\widetilde{\varepsilon}(\nu) = \varepsilon'(\nu) - i\varepsilon''(\nu)$ and $\widetilde{\varepsilon}(\nu) = \widetilde{n}_2^2(\nu)$). At the same time, the spectroscopy data is not readily analyzed by such representation. Much more convenient basis for data discussion and analysis could be Argand diagrams (complex plane locus or Cole-Cole diagrams) [37,38], in which the imaginary part of complex dielectric constant is plotted against the real one, $\varepsilon''(\varepsilon')$, and each point being characteristic of certain electromagnetic wave frequency. Since the material parameters of *in vivo* biological tissue have a Debye-like character [39,40], it is convenient to represent the THz dielectric characteristics of *in vivo* tissue using the Cole-Cole diagram representation. The equation, describing the frequency dependencies of ε' and ε'' for Debye model, has the following form

$$\varepsilon'(\nu) = \varepsilon_\infty + \frac{\varepsilon_0 - \varepsilon_\infty}{1 + (2\pi\nu\tau_0)^2}, \qquad \varepsilon''(\nu) = \frac{2\pi\nu\tau_0(\varepsilon_0 - \varepsilon_\infty)}{1 + (2\pi\nu\tau_0)^2}, \tag{8}$$

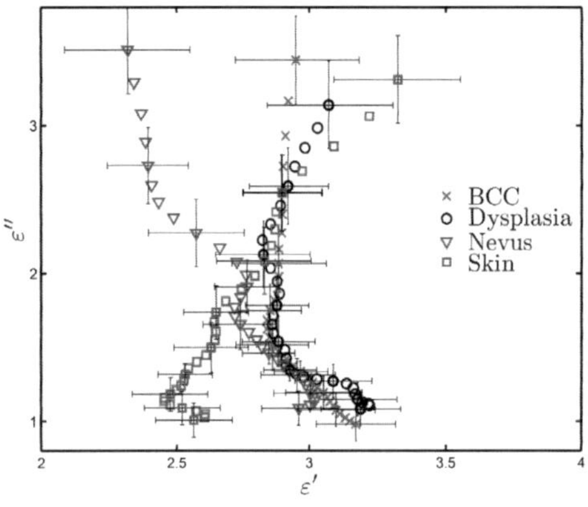

Figure 3. Cole-Cole diagrams of healthy skin and pathology *in vivo*: basal cell carcinoma (×), dysplastic pigmentary nevus (◯), non-dysplastic pigmentary nevus (▽), and healthy skin (□).

where ε_0, ε_∞ are model parameters, and τ_0 is a medium relaxation time. The locus of equation (8) in complex plane representation of experimental results is a semicircle with its center on the real axis, ε', and interceptions with this axis at $\varepsilon' = \varepsilon_0$ and $\varepsilon' = \varepsilon_\infty$ In case of multiple-Debye-model describing the sample dielectric characteristics, a set of semicircles exist in the complex plane.

We are applying described Cole-Cole representation to all results of the figure 3. Due to the boundedness of reliable spectral range the Cole-Cole diagrams represent only parts of semicircles; but the experimental data significantly differs for all studied tissues, providing us the ability for principle components selection based on the complex plane locus. We are able to find the differences between basal cell carcinoma and healthy skin, as well as the differences between the dysplastic and non-dysplastic pigmentary nevi.

Dysplastic pigmentary nevus is a melanoma precursor, thus, observed result shows the ability for early melanoma diagnosis, and it is an important problem of dermatology and oncology, since melanoma is the most dangerous cancer of skin [34,35,41,42]. Described results of *in vivo* spectroscopy allow to conclude that THz spectroscopy could become a powerful instrument for diagnosis of skin diseases, including, skin cancer diagnosis.

4. Conclusions

In the present paper we have measured the THz material parameters and THz dielectric properties of healthy skin and pathology *in vivo* by by using THz pulsed spectroscopy. The material parameters and dielectric characteristics of basal cell carcinoma, dysplastic and non-dysplastic pigmentary nevi, and healthy skin have been examined. We have found significant differences between THz characteristics of healthy skin and pathology, and this allows to conclude, that THz pulsed spectroscopy could become an effective tool for skin neoplasm diagnosis.

Acknowledgements

This work was supported by the Russian Scientific Foundation, Project # 14-15-00758.

References

[1] Lee Y-S 2009 *Principles of Terahertz Science and Technology* (New York: Springer)
[2] Auston D H 1975 Appl. Phys. Lett. **26** 101–03
[3] Duvillaret L, Garet F, Coutaz J-L 1999 *Appl. Opt.* **38** 409–15
[4] Dorney T D, Baraniuk R G, Mittleman D M 2001 *J. Opt. Soc. Am. A* **18** 1562–71
[5] Pupeza I, Wilk R, Koch M 2007 *Opt. Exp.* **15** 4335–50

[6] Huang S, Ashworth P C, Kan K W C, Chen Y, Wallace V P, Zhang Y-T, and Pickwell-MacPherson E 2009 *Opt. Exp.* **17** 3848–54
[7] Balakrishnan J, Fischer B M, Abbott D 2010 *Opt. Commun.* **283** 2301–7
[8] Scheller M 2011 *Opt. Exp.* **19** 10647–55
[9] Zaytsev K I, Gavdush A A, Lebedev S P, Yurchenko S O 2014 *J. Phys. Conf. Ser.* **486** 012018
[10] Kruger M, Funkner S, Brundermann E, Havenith M 2011 *J. Infrared, Millimeter, and Terahertz Waves* **32** 699–715
[11] Zaytsev K I, Gavdush A A, Karasik V E, Alekhnovich V I, Nosov P A, Lazarev V A, Reshetov I V, Yurchenko S O 2014 *J. Appl. Phys.* **115** 193105
[12] Mittleman D M, Hunsche S, Boivin L, Nuss M C 1997 *Opt. Lett.* **22** 904–6
[13] Chen Y, Huang S, and Pickwell-MacPherson E 2010 *Opt. Exp.* **18** 1177–90
[14] Zaytsev K I, Karasik V E, Fokina I N, Alekhnovich V I 2013 *Opt. Eng.* **52** 068203
[15] Zaytsev K I, Chernomyrdin N V, Alekhnovich V I 2014 *J. Phys. Conf. Ser.* **486** 012010
[16] Woodward R M, Wallace V P, Pye R J, Cole B E, Arnone D D, Lineld E H, and Pepper M 2003 *J. Investigative Dermatology* **120** 72–78
[17] Pickwell E and Wallace V P 2006 *J. Phys. D: Appl. Phys.* **39** R301-R310
[18] Reid C B, Fitzgerald A, Reese G, Goldin R, Tekkis P, OKelly P S, Pickwell-MacPherson E, Gibson A P, and Wallace V P 2011 *Physics in Medicine and Biology* **56** 4333-53
[19] Doradla P, Alavi K, Joseph C, and Giles R 2013 *J. Biomed. Opt.* **18** 090504
[20] Fitzgerald A J, Wallace V P, Jimenez-Linan M, Bobrow L, Pye R J, Purushotham A D, Arnone D D 2006 *Radiology* **239** 533–40
[21] Ashworth P C, Pickwell-MacPherson E, Provenzano E, Pinder S E, Purushotham A D, Pepper M, and Wallace V P 2009 *Opt. Exp.* **17** 12444–54
[22] Bennett D B, Taylor Z D, Tewari P, Singh R S, Culjat M O, Grundfest W S, Sassoon D J, Johnson R D, Hubschman J-P, and Brown E R 2011 *J. Biomed. Opt.* **16** 057003
[23] Pickwell E, Wallace V P, Cole B E, Ali S, Longbottom C, Lynch R J M, Pepper M 2007 *Caries Research* **41** 49-55
[24] Sy S, Huang S, Wang Y-X J, Yu J, Ahuja A T, Zhang Y-T, and Pickwell-MacPherson E 2010 *Physics in Medicine and Biology* **55** 7587-96
[25] Jeong K, Huh Y-M, Kim S-H, Park Y, Son J-H, Oh S J, Suh J-S 2013 *J. Biomed. Opt.* **18** 107008
[26] Hassan Arbab M, Dickey T C, Winebrenner D P, Chen A, Klein M B, and Mourad P D 2011 *Biomed. Opt. Exp.* **2** 2339–47
[27] Tewari P, Kealey C P, Bennett D B, Bajwa N, Barnett K S, Singh R S, Culjat M O, Stojadinovic A, Grundfest W S, and Taylor Z D 2012 *J. Biomed. Opt.* **17** 040503
[28] Joseph C S, Yaroslavsky A N, Al-Arashi M, Goyette T M, Dickinson J C, Gatesman A J, Soper B W, Forgione C M, Horgan T M, Ehasz E J, Giles R H, and Nixon W E 2009 *Proc. SPIE* **7215** 72150I
[29] Mazhorova A, Markov A, Ng A, Chinnappan R, Skorobogata O, Zourob M, and Skorobogatiy M 2012 *Opt. Exp.* **20** 5344–55
[30] Son J-H 2013 *Nanotechnology* **24** 214001
[31] Walther M, Plochocka P, Fischer B, Helm H, Uhd Jepsen P 2002 *Biopolimers* **67** 310–13
[32] George P A, Hui W, Rana F, Hawkins B G, Smith A E, Kirby B J 2008 *Opt. Exp.* **16** 1577–82
[33] Oh S J, Choi J, Maeng I, Park J Y, Lee K, Huh Y-M, Suh J-S, Haam S, and Son J-H 2011 *Opt. Exp.* **19** 4009–16
[34] Grant-Kels J M 2013 *Expert Review of Dermatology* **8** 95-97
[35] Rees J 2013 *Expert Review of Dermatology* **8** 441–41
[36] Wilmink G J, Ibey B L, Tongue T, Schulkin B, Laman N, Peralta X G, Roth C C, Cerna C Z, Rivest B D, Grundt J E, and Roach W P 2011 *J. Biomed. Opt.* **16** 047006
[37] Cole K S, and Cole R H 1941 *J. Chem. Phys.* **9** 341–50
[38] Cole K S, and Cole R H 1942 *J. Chem. Phys.* **10** 91–105
[39] Pickwell E, Cole B E, Fitzgerald A J, Wallace V P, and Pepper M 2004 *Appl. Phys. Lett.* **84** 2190–92
[40] Yurchenko S O, and Zaytsev K I 2014 *J. Appl. Phys.* **116** 113508
[41] Reshetov I V, Potekaev N N, Arutyunyan L S, Zaletaev D V, Matorin O V, Kudrin K G 2009 *Russian Oncological Journal* **5** 54–56
[42] Kemeny L V, Kurgyis Z, Polyanka H, Dittmar T, Kemeny L, and Nemeth I B 2014 *J. Investigative Dermatology* **134** S90-S97

Combined holographic optical elements for multicolor holographic screens and indicators

Sergey B Odinikov*, Alexey S Kuznetsov, Vasiliy V Kolyuchkin, Ekaterina A Drozdova, Artem B Solomashenko**
Bauman Moscow State Technical University, 105005, 2nd Baumanskaya str. 5, Moscow, Russia

E-mail: *odinokov@bmstu.ru, **art_s87@mail.ru

Abstract. A combination of holographic and diffractive optical elements (HOE-DOEs) for multicolor signs-symbolic information display system, screens and indicators is described. These elements combine the functions of high-efficiency four-level diffractive optical elements with spectral plasmon filters having a variable bandwidth dependent on the angle of incidence. The main parameters of such HOE-DOEs were determined using the Fourier modes method. The theoretical results show that a four-level diffraction grating and a plasmon filter can be combined based on their spectral-angular characteristics, opening up the possibility for their use in display systems.

1. Introduction

Currently, one of the most promising practical applications of flat optics is the development of holographic and diffractive optical elements (HOE-DOEs) with binary and multi-level surface relief, as well as their integration into optical devices. HOE-DOEs with light guide substrates are used in miniature display systems [1-6]. With these combined elements, the weight and size of display and visualization systems can be significantly reduced [3,4], which allows for their use in new applications, especially in automotive and aviation systems [2,4,5], where the display is superimposed on real world imagery.

The use of holographic indicators in different climatic conditions requires the selection of special photosensitive materials for HOE-DOEs. These elements may need to be physically protected or recorded directly into the glass [4]. There is a significant problem with the low diffraction efficiency of these devices [1-3]. Improved efficiency would allow for the use of less powerful illumination source including the possibility of using OLED-displays. In order to project multi-colored images, HOE-DOEs need to be designed with the specified spectral and angular selectivity. The development of a new type of combined HOE-DOEs with increased diffraction efficiency and spectral-angular selectivity for different photosensitive materials and glass will facilitate the development of a new generation of signs-symbolic information displays, based on holographic optical elements and light guide substrates.

2. Combined holographic optical element scheme

The main component of a holographic indicator is a signs-symbolic information output device, which can be mounted directly on the head of pilots, soldiers, vehicle riders, etc. These devices include a glass substrate [1-2], in which radiation can propagate with total internal reflection, and diffraction

Content from this work may be used under the terms of the Creative Commons Attribution 3.0 licence. Any further distribution of this work must maintain attribution to the author(s) and the title of the work, journal citation and DOI.
Published under licence by IOP Publishing Ltd

gratings (DG) at the input and output [1-2]. A major disadvantage of these relief-phase gratings is their low diffraction efficiency and their location on a layer of photosensitive material, which may not be suitable for operational conditions. Therefore, either this material needs to be protected or the grating must be fabricated directly in the glass with maximum diffraction efficiency, i.e. multi-level structures rather than binary structures are required, as shown in Figure 1.

HOE-DOEs can perform with increased efficiency for signs-symbolic information display DGs by using a few quantization levels of the surface microrelief. To generate multicolor images by holographic indicators, spectral filters must be used for selecting the incident wavelengths.

The most promising method for spectral selection when the source is white or polychromatic light is the use of plasmonic effects in metal-dielectric thin films [7-8]. This enables the creation of broadband and narrow band spectral filters with adjustable bandwidth by changing the angle of incidence of the input light. The use of these spectral filters in the new generation of miniature devices and data indication systems provides a high spectral-angular selectivity of transmitted or reflected optical radiation, which allows for the display of multicolor images.

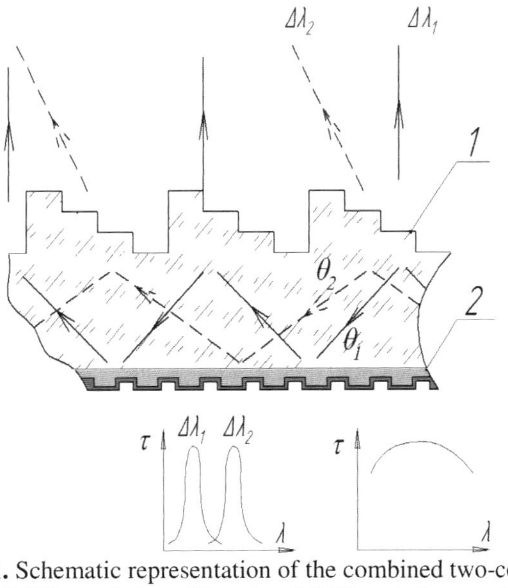

Figure 1. Schematic representation of the combined two-component HOE-DOE consisting of four-level DG1 and plasmon filter 2.

A combined two-component HOE-DOE configuration is shown in Figure 1, where a phase four-level element outputs the light, and a metallized diffraction grating performs the spectral selection. The rectangular metal diffraction grating has a transmission spectrum of the zero diffraction order, which depends on the angle of incidence. Depending on the angle of incidence of the polychromatic radiation, the plasmon filter selects which wavelengths are directed to a four-level HOE-DOE. A two- to four-level DG is used as the HOE-DOE.

3. Theoretical studies

The diffraction efficiency of any diffractive structure can be determined by Maxwell's equations. Numerous studies are devoted to finding a rigorous solution for electromagnetic diffraction by periodic structures. In this paper, the Fourier modes method is used to determine the parameters of the DGs. This is one of the most common methods used [9-10]. It is implemented using MATLAB software. Using his software, the diffraction of a plane wave on a DG is solved for various structures and the spectral characteristics and energy parameters of different diffraction orders are calculated.

In this case, the following model is considered. Light propagates by total internal reflection within the glass substrate that has a refractive index of 1.51, as shown in Figure 1. The grating has a period d= 490 nm [3], and its surface microrelief consists of 2–4 levels. The radiation has wavelengths of 440, 532 and 635 nm and is TE- polarized. Light from the first diffraction order is diffracted out of the substrate by DG 1.

Basic parameters of the HOE-DOEs are calculated by the software. The relief depth of the four-level grating is 350 nm. The diffraction efficiency and spectral characteristics corresponding to different wavelengths and incidence angles are shown in figure 2.

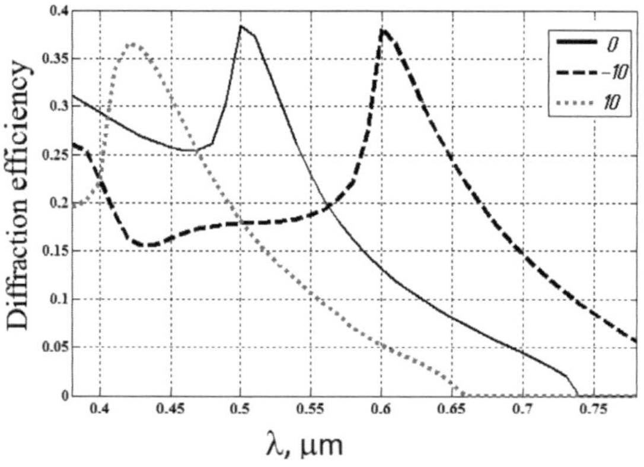

Figure 2. Diffraction efficiency of four-level profile diffraction grating corresponding to wavelength of incident radiation. The angles of incidence are -10°, 0°, and 10°.

A plasmon filter [7-8] may be used for spectral selection. It consists of a glass substrate and a photoresist layer with a meander grating coated by a layer of silver and SiO_2. The meander structure can be produced in the photoresist or glass layer by lithography and ion-plasma etching technology. The features of the resulting structure are investigated.

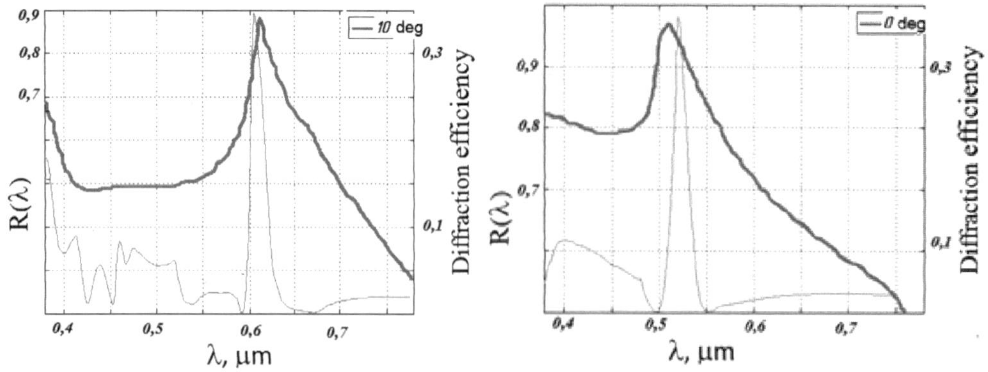

Figure 3. Diffraction efficiency of the four-level diffraction grating at incidence angles of 10° and 0° (dark lines) and radiation spectrum R(λ) reflected from the plasmon filter (light lines), corresponding to wavelength of incident radiation.

The optimal parameters of a spectral filter, which is a part of the combined HOE-DOE, are a meander structure with a period of 500 nm on a glass substrate with a refractive index n = 1.51, a resist layer with a thickness of 40 nm, a metallic silver layer of thickness 30 nm, and a SiO_2 layer of thickness 20 nm. Under these parameters, the spectral peak width is less than 20 nm and a strong correlation of the reflected radiation wavelength with its angle of incidence is observed, which suggests the possibility of obtaining a multicolor image.

The graphs in Figure 3, calculated by Fourier modes method, show that the two elements (four-level DG and plasmon filter) are combined by their spectral-angular characteristics, and can thus work together, which indicates the possibility of obtaining a combined HOE-DOE.

4. Conclusion

The combination of four-level diffractive optical elements having high diffraction efficiency with spectral filters that have a bandwidth that varies with the angle of incidence inside a single HOE-DOE has been explored. This is a new direction for the development of miniaturized data indication systems. To analyze these elements, Fourier modes method is used for solving Maxwell's equations. It involves the implementation of special software to solve the direct diffraction of a plane electromagnetic wave by a meander structure. The theoretical results suggest the possibility of using this combined HOE-DOE for displays, screens, and indicators showing signs-symbolic information.

Acknowledgments

The research was carried out at BMSTU. This work was supported by a grant from the Ministry of Education and Science of the Russian Federation # 14.574.21.0066 (Project ID RFMEFI57414X0066).

References

[1] Eisen L, Meyklyar M, Golub M, Friesem A, Gurwich I, Weiss V 2006 *Appl. Opt.* **45** 4005-11
[2] Eisen L, Meyklyar M, Friesem A, Gurwich I, Weiss V 2003 *Proc. SPIE* **5182** 205-21
[3] Odinokov S, Markin V, Kuznetsov A, Drozdova E, Solomashenko A 2013 *Proc. SPIE* **8776** 87760Q
[4] Cameron A. 2012 *Proc. SPIE* **8383** 83830E
[5] Shechter R, Amitai Y, Friesem A 2002 *Appl. Opt.* **41** 1236-40
[6] Yan Z, Li W, Zhou Y, Kang M, Zheng Z 2011 *Opt. Eng.* **50**(9) 094001
[7] Fu L, Schweizer H, Weiss T, Giessen H 2009 *J. Opt. Soc. Am.* B. **26** B111–B119
[8] Ozaki M, Kato J, Kawata S 2011 *Science* **332** (6026) 218-20
[9] Moharam M G, Grann E B, Pommet D A, Gaylord T K 1995 *J. Opt. Soc. Am. A.* **12** 1068–76
[10] Moharam M G, Grann E B, Pommet D A, Gaylord T K 1995 *J. Opt. Soc. Am. A.* **12** 1077–86

Problem of light scattering in complex media

Egor V Yakovlev*, Irina N Dolganova, Kirill I Zaytsev, Tauso I Kuliev, Nikita P Kruchkov and Stanislav O Yurchenko
Bauman Moscow State Technical University, 2nd Baumanskaya str., 5, 105005, Moscow, Russian Federation

E-mail: *yakov.egor@gmail.com

Abstract. This work includes discussion about different types of interaction between electromagnetic radiation and various materials. Field reorganization in ordered media is considered, also impact of scatterers' size and distance between them on the scattering process is discussed. Phase diagram for scattering regimes in complex media is suggested.

1. Introduction

In the end of XIX century the influence of radiation on the medium properties was studied systematically due to the photoelectric effect. The internal and external photoelectric effect is the result of interaction between light and medium, which leads to electrons fly out the medium and to the ionization. Thus, even weakly conductive media become conductors and their properties change significantly in electromagnetic fields. By the example of photoelectric effect the importance of interaction between light and medium becomes more evident.

Interaction between radiation and materials can be divided into two groups: absorption and scattering. A lot of publications concern absorption of electromagnetic radiation [1], this problem is well studied. Depending on the radiation range, the interaction between electromagnetic radiation and medium can be caused by absorption due to rotational and vibrational terms of molecules (microwave, terahertz, infrared ranges), or due to transitions between molecule electron levels (visual range), ionization by UV and X-rays, and also due to Compton scattering. In comparison with absorption, scattering of radiation is weakly studied.

2. Electromagnetic scattering in ordered medium

Propagation of radiation even in homogeneous media is rather complex. While radiation penetrate the material, it scatters and shifts in spectrum, which depends on the spatial orientation of scattered radiation.

But in fact, the complexity of radiation interaction with medium increases with the presence of structural material (also periodical). Thus, collective effects of coherent interaction (multi-ray interference) should be considered while studying scattering, which are so strong in ordered media, such as metameterials, photonic crystals, and composite materials [2].

Structure effect is more significant in the electromagnetic range, where wavelength is comparable to the size and the period of the structure. Until recently it was believed that spatial periodicity has little effect on scattering in such media, but last studies show the opposite.

For structures with long-range correlations (which means the long-range correlations of properties), the weak processes (for example, scattering of radiation), the results of which add coherently with each other, become decisive. In this case electromagnetic field is redistributed inside the material, also

Content from this work may be used under the terms of the Creative Commons Attribution 3.0 licence. Any further distribution of this work must maintain attribution to the author(s) and the title of the work, journal citation and DOI.
Published under licence by IOP Publishing Ltd

the field localization occurs [3-5]. Today electromagnetic field interaction with optical materials, which have long-range correlations of optical properties also when multiple scattering is taken into account, is a complicated and poorly understood problem. Therefore, in order to study terahertz (THz) waves interaction with quasi-crystal media and metamaterials, the numerical methods are widely used. The approaches, which account the structure factor, are the most important in this case.

3. Approaches to describe the radiation scattering

Nowadays there are various methods to account the material structure for studying the radiation-material interactions. Figure 1 shows the basic approaches used in different ranges of electromagnetic spectrum.

Historically, radiation transfer theory was derived first. In different modifications it is used for wide range of electromagnetic waves: from microwaves to X-rays [6]. This method is based on the radiation transfer equation, which should be completed by the appropriate "closed" equations for the scattering phase function (scattering indicatrix). Thus, in this sense, the radiation transfer theory is not completed and needs additional assumptions about the character of radiation propagation in the media.

Along with the phenomenological method, the radiation transfer equation can be derived from Maxwell equations.

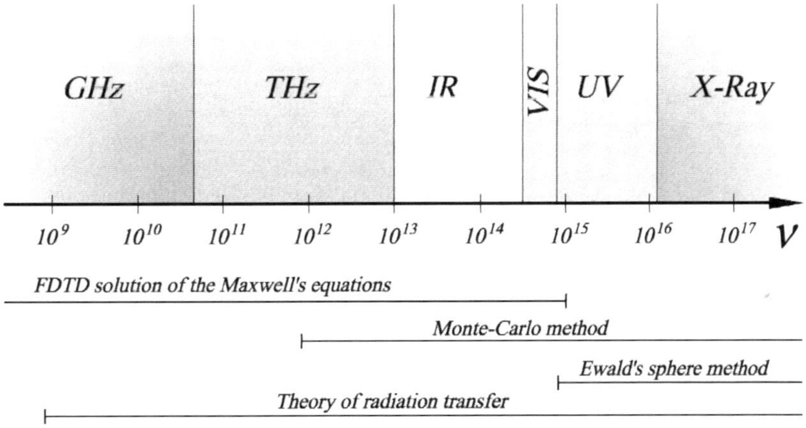

Figure 1. Application fields of the basic methods for description of the field propagation within the structures with long-range correlations.

In radiation transfer theory in case of rarified disordered media without short- and long-range correlations between spatial geometry and material properties, various *a priori* approximations estimating single scattering phase function are used. The contribution of multiple scattering is assumed to be negligible.

With the growth of packing density, in other words, with the decreasing of the distance between scatterers, multiple scattering shows a significant impact. Thus, uncertainty appears even in the *a priory* choice of scattering phase function for complex media, and especially for media with long-range correlations of properties. It is necessary to use basic theoretical principles for radiation transfer computations. Generally, the most frequently used methods for this purpose are computational electrodynamics [2,3-5,7], Monte Carlo numerical simulation considering the quantum nature of light [6,8,9], and the Ewald concept [10,11], which was developed for X-rays transfer and spread on the optical range.

Using these methods, it is unnecessary to get *a priory* estimation of the scattering character. Moreover, it is possible to find regularities in radiation transfer in different media. Figure 2 illustrates the discussed problem of scattering influence on the radiation transfer character in medium. The proposed picture from figure 2 is the phase diagram of scattering regimes developed for complex media.

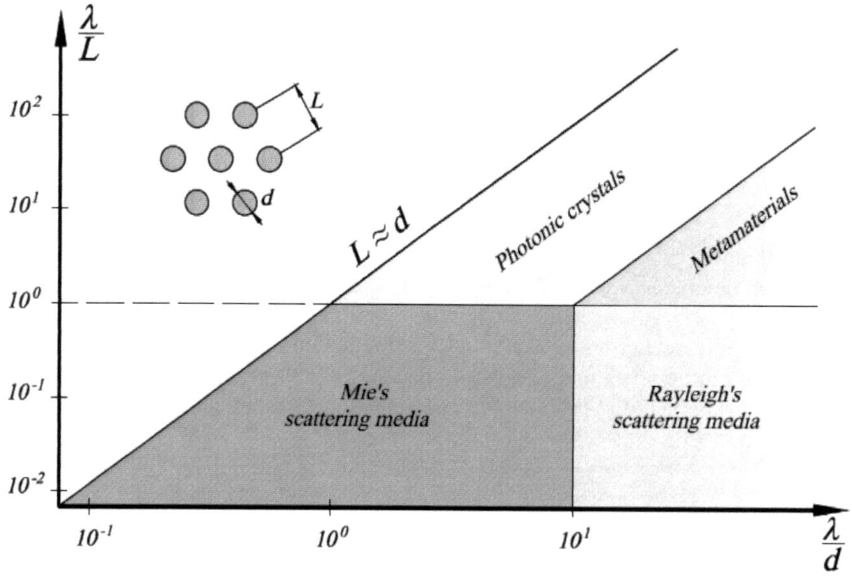

Figure 2. Phase diagram for possible scattering regimes.

When the distance between scattering particles or the correlation length of structure is much larger than wavelength, scattering has the incoherent character. Such process is well described by the Mie and Rayleigh scattering theories, thus, no difficulties can be found in this case.

Otherwise, description of the coherent scattering is more complicated. Depending on the relation between wavelength and correlation length, coherent scattering may be described by the photonic crystal theory or the theory of metamaterials. These approaches were developed relatively recently and today attract significant attention.

It should be noticed, that methods of radiation transfer theory is ineffective in case of photonic crystal and quasi-crystal media, metamaterials, because of uncertainty of scattering phase function used in order to complete radiation transfer equation. In the same time, using methods of computational electrodynamics, the visualization of field redistribution inside the scattering material becomes possible, as well as determining of the scattering phase functions and establishment of regularities for radiation transfer in complex media.

4. Conclusion

Possible types of interaction between electromagnetic radiation and scattering medium are analysed in dependence of different radiation frequency ranges. Also the impact of scattering structure on the radiation propagation is discussed. Phase diagram for scattering regimes in complex media was suggested for generalization of different scattering theories and approaches. Using this diagram, all scattering materials can be classified unambiguously and the radiation transfer can be considered in a new way.

Acknowledgement

This work was supported by a grant from the Ministry of Education and Science of Russian Federation (№ 14.577.21.0095).

References

[1] Murphy E J, Morgan S O 2013 *Bell System Technical Journal* **16**(4) 493–512

Journal of Physics: Conference Series **584** (2015) 012025 doi:10.1088/1742-6596/584/1/012025

[2] Yakovlev E V, Zaytsev K I, Fokina I N, Karasik V E, Yurchenko S O 2014 *J. Phys. Conf. Ser.* **486** 012008
[3] Zaytsev K I, Yurchenko S O 2014 *Appl. Phys. Lett.* **105** 051902
[4] Zaytsev K I, Katyba G M, Yakovlev E V, Gorelik V S, Yurchenko S O 2014 *J. Appl. Phys.* **115** 213505
[5] Zaytsev K I, Gorelik V S, Katyba G M, Yurchenko S O 2014 *J. Phys. Conf. Ser.* **541** 012072
[6] Collins D G, Blattner W G, Wells M B, Horak H G 1972 *Appl. Opt.* **11**(11) 2684-2696
[7] Yee K S 1966 *IEEE Trans. Antennas Propagat.* **AP-14**(3) 302-307
[8] Mikhailov G A, Tracheva N V, Ukhinov S A 2008 *Russ. J. Numer. Anal. Math. Modelling.* **23**(3) 239-250
[9] Fokina I N, Karasik V E, Orlov V M, Budak V P 2014 J. Quantitative Spectroscopy and Radiative Transfer **149** 108-116
[10] Litzman O, Mikulík P, Dub P 1966 *J. Phys. (UK)* **8** 4709-4725
[11] Mikulík P, Baumbach T 1998 *Physica B* **248** 381-386

Lowering of stimulated Raman scattering threshold as a result of light capture

V S Gorelik[1], A M Negriyko[2], V A Orlovich[3], P P Sverbil[1], N V Tcherniega[1], A I Vodchits[3], Y P Voinov[1] and L I Zlobina[1]

[1] P.N. Lebedev Physical Institute of the Russian Academy of Sciences, Russia

[2] Institute of Physics of the National Academy of Sciences of Ukraine

[3] B.I. Stepanov Institute of Physics, National Academy of Sciences of Belarus

E-mail: gorelik@sci.lebedev.ru

Abstract. Stimulated Raman Scattering in globular photonic crystals and globular photonic glasses at different diameters of globules (250 - 400 nm) with embedded molecular liquids is studied under excitation by nanosecond or picoseconds laser pulses. Substantial decrease of Stimulated Raman Scattering threshold was observed. Such phenomenon was explained as the result of laser radiation field increase in globular photonic structures due to photonic density of states enhancement near the edges of photonic stop bands of photonic crystals and due to Mie resonance or whispering gallery modes effect revealing in photonic glasses. Stimulated Raman Scattering threshold lowering as a result of light capture in globular photonic crystals and photonic glasses opens the way to new efficient laser sources created on the base of composite globular photonic structures. Experimental data on spectra of Stimulated Raman Scattering in light and heavy waters are presented. As sources of exciting light the powerful ultra short solid state laser pulses with 532.0 nm wavelength and giant pulses of Ruby laser (694.3 nm) have been used. Several Stokes and anti-Stokes satellites were observed. Libration modes have been excited and resulted in some additional Raman bands at low frequency region and also as combining tones.

1. Introduction and background

Spontaneous Raman Scattering (RS) effect is widely used for obtaining of information on vibration spectra of solids and liquids [1-3]. However RS spectra are difficult for registration because of very small cross-section (10^{-28} cm^2) and accordingly very small value of intensity in comparison with exciting emission one (10^{-6} -10^{-8}) . Stimulated Raman Scattering (SRS) effect has been observed for a number of liquids and solids. In this case the scattering intensity is comparable with exciting emission one. However, as a rule, SRS threshold is high enough (10-100 GW/cm^2) and is close to destruction threshold of condensed media. So there is a problem of searching the ways of SRS threshold lowering. Before [4-6] there were made predictions of lasing- threshold lowering in strongly scattering condensed disordered systems (powders or composites).

New possibilities in SRS investigations have been opened in connection with composite technologies development [7-9]. Very promising materials for SRS threshold lowering are highly scattering composite media. The examples of such type media are artificial opals, consisting from SiO_2 – globules, packed into cubic face centered lattice. Crystalline lattice constant of artificial opal is close to visible radiation wavelength (300-600 nm). If the globules with the same diameter are disordered in the space, so called photonic glass is formed. Between the globules in initial globular photonic

Content from this work may be used under the terms of the Creative Commons Attribution 3.0 licence. Any further distribution of this work must maintain attribution to the author(s) and the title of the work, journal citation and DOI.
Published under licence by IOP Publishing Ltd

structures there is free space; so this space may be filled by some Raman active substances. Modern technology permits also to construct spherical photonic structures from amorphous SiO_2 – units with diameters of micro- or millimeter scales. Thus globular photonic structures with Raman active embedding substances between globules are perspective objects for SRS threshold lowering. Other examples of globular photonic structures are spherical particles of micrometer scale, consisting from thin amorphous quartz spherical walls with air inside. It is interesting also to use the ordered in space amorphous quartz balls of millimeter size for realizing strongly light scattering structures. In this work we report the results of experimental investigations of SRS in some homogeneous liquids, including light (H_2O) and heavy (D_2O) waters, and embedded into pores of photonic structures liquid media.

2. Experimental technique

SRS was excited whether by giant (10 ns) pulses of Ruby laser or by ultra short (60 ps) pulses of the second harmonic radiation of Nd:YAG laser. Experimental scheme for SRS spectra recording under ultra short laser pulses excitation is presented at figure 1a. In this case exciting emission corresponded to the second harmonic (532.0 nm) radiation of Nd:YAG laser (1064.0 nm).

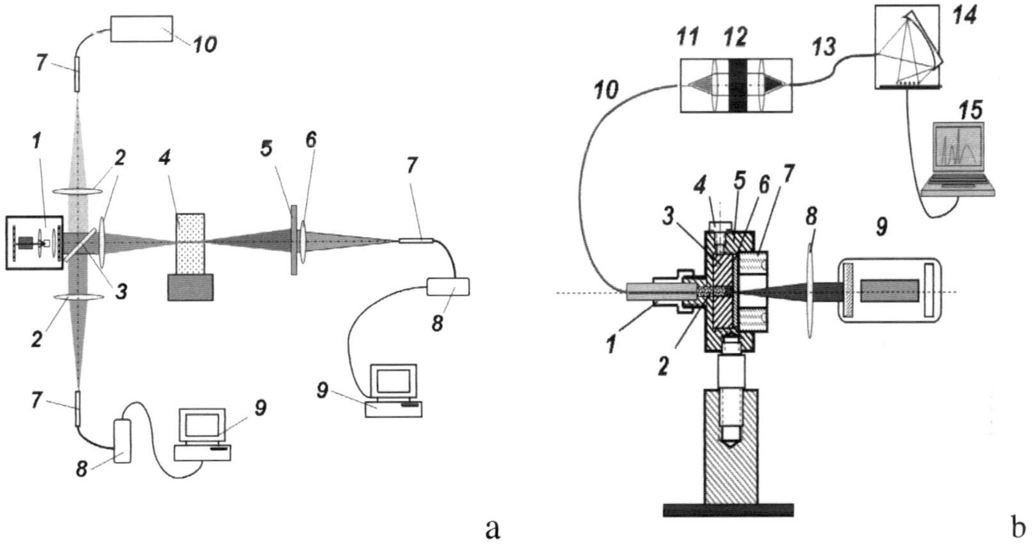

Figure 1. Experimental setup for SRS spectra , excited by powerful laser pulses (a) with the help of ultra short pulses: 1-laser, 2,6-lens, 3-qurtz plate, 4-sample, 5-filter, 7-fiber, 8- minispectrograph, 9-computer, 10 - detector; b- with the help of giant laser pulses: 1-fiber holder, 2-photonic crystal (glass), 3,6-plates, 4,7-covering cylinders, 5-cuvette, 8-lens, 9-Ruby laser, 10,13-fibers,11- -filter, 14-minispectrograph, 15-computer.

The exciting laser (1) emission was focused into cuvette with investigated liquid or to photonic structure (crystals or glasses), pores of which were filled with investigated liquid. Experimental setup permitted us to register the Raman scattered light spectrum in the forward and backward directions with the help of minispectrograph 8 and computer 9. Detector 10 allows to measure the laser intensity. The similar scheme has been used for the case of giant pulses of Ruby laser (694.3 nm) SRS excitation (see figure 1b). We have used picoseconds YAG:Nd^{3+} laser with the following parameters: energy was in the range 0.1 - 35 mJ at 532 nm wavelength, pulse duration - 60 ps, pulse repetition rate - 20 Hz. Laser beam was focused into the quartz cell with water with the help of large focus lens (200 mm). After suppression of exciting emission by prism, diffraction lattice or absorbing filter, the secondary emission from water sample was directed to the fiber tip. SRS spectra were registered by FSD-8 minispectrograph, having CCD-detector with 200 – 1000 nm spectral range.

3. Experimental results and discussion

Figure 2 illustrates SRS-spectra for nitrobenzene filled photonic crystal, excited by giant single pulses of red (694.3 nm) emission of Ruby laser. We have seen two Stokes components at 766 and 854 nm correspondingly. The diameters of quartz globules of artificial opal in this case were 310 nm. So the spectral positions of exciting line (694.3 nm) and the first Raman satellites (766 nm) were close to the edges of the first photonic stop-band positions. At shorter wavelengths (see figure 2) there was a number of satellites, which we attribute to Two-Photon Excited Photoluminescence. We have established that the threshold of SRS-processes in nitrobenzene filled photonic crystal was essentially lower than in pure nitrobenzene as a result of light capture.

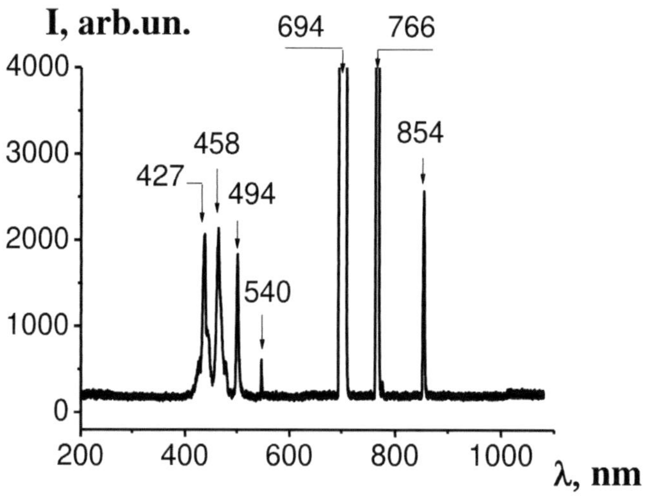

Figure 2. SRS spectra of nitrobenzene-filled photonic crystal, excited by giant pulses of Ruby laser.

Figure 3 illustrates SRS-spectra of light water filled artificial opal, excited by second harmonic (532.0 nm) of picoseconds YAG: Nd^{3+} - laser. At figure 3a we can see exciting line (532.0 nm) and the first Stokes component of water at 650 nm (3339 cm^{-1}). In this case the diameter of quartz globules of artificial opal was 260 nm. Accordingly we have a proximity of laser line spectral position (532.0 nm) and the first Stokes position(650.0 nm) to the edges of corresponding stop- band of photonic crystal. The SRS emission of light water or ethanol, excited by powerful picoseconds pulses, was well seen as red emission with naked eye. In this case SRS beams propagated in forward and backward directions. SRS spectrum of light water consists of Stokes (see figure 3) and anti-Stokes satellites.

Some shoulder appeared near anti-Stokes fundamental $\nu_1(A_1)$=3424 cm^{-1} mode spectral position. At small exciting energy (~1mJ), before threshold achieving, the observed Stokes SRS spectrum had the shape of wide band with several peaks, corresponding to fundamental $\nu_1(A_1)$ mode (figure 3c). After increasing of the exciting energy (to 10mJ) Raman intensity of this band increased and its width essentially decreased (figure 3d). With increasing of exciting emission power, spectral position of SRS peak intensity shifted from 3186 up to 3424 cm^{-1}. At backward scattering geometry, at low frequency region there were two wide additional satellites (Stokes and anti Stokes) at 780 cm^{-1}, attributed to librational mode $\nu_{Lib.}$. Weak maximum at 1650 cm^{-1} corresponds to $\nu_2(A_1)$ fundamental mode.

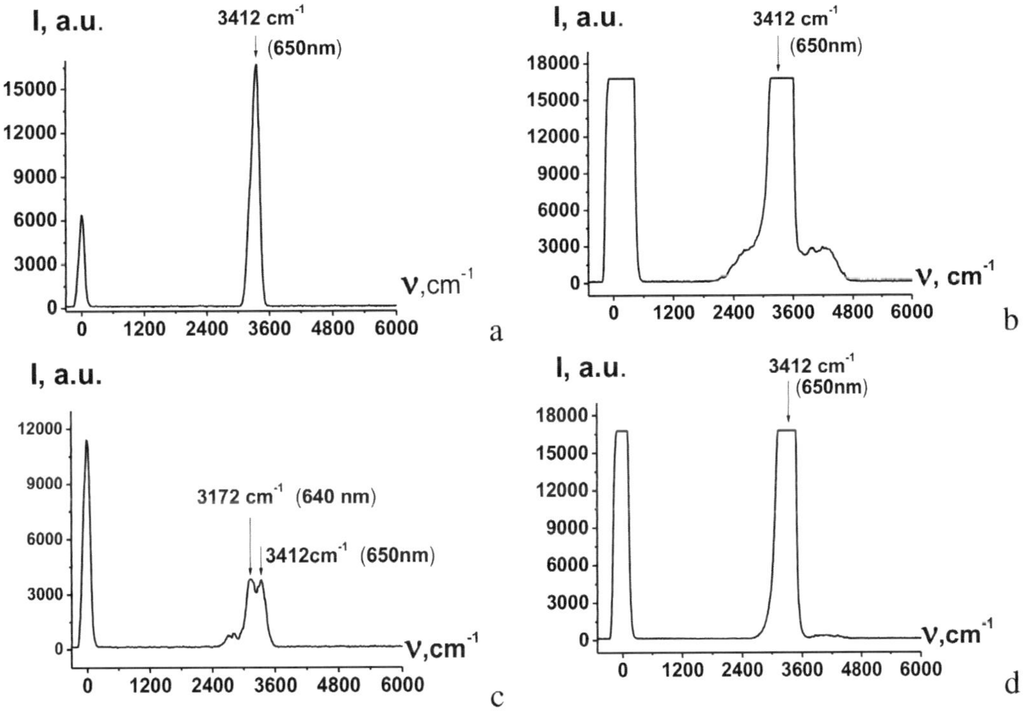

Figure 3. The shape of forward SRS spectra recorded near the surface of artificial opal filled by water at lower (a) and higher (b) values of exciting laser power and in pure water (c, d), correspondingly.

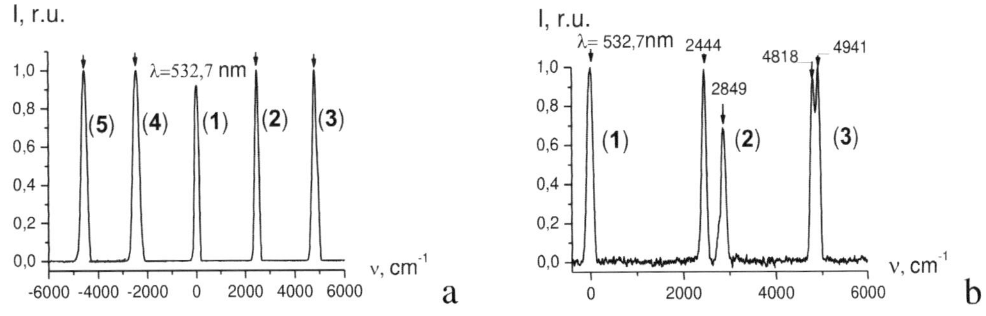

Figure 4. The shape of the normalized heavy water SRS-spectrum under picosecond excitation; a- Stokes(2,3) and antiStokes(4,5) satellites;(1)-exciting line; b- the shape of the first and second Stokes bands at higher power of exciting radiation.

Figure 4 illustrates registered SRS spectra of heavy water under picoseconds excitation for backward scattering geometry. In SRS spectrum of heavy water the most intensive line with frequency $\nu_1(A_1)=2448$ cm^{-1} has been observed. We can see in figure 4a two Stokes and two anti-Stokes satellites. The shape of the first and second Stokes bands under higher intensity of exciting emission is illustrated in figure 4b. We can conclude, there at the first Stokes region there are two distinct lines, one of which corresponds to $\nu_1(A_1)$ fundamental mode and another – to combinational states: $\nu_1(A_1)$ $+\nu_{Lib}$ of heavy water. Accordingly some additional structure was at the second Stokes band (see figure 4b).

4. Conclusions

As it was predicted by theory [10-12], the spontaneous Raman scattering cross sections in liquids filled photonic crystals should depend on the photonic stop-band spectral position due to the sharp increase of photonic density of states near the edge of photonic stop-band. Therefore, SRS threshold in liquids filled artificial opals should decrease as a result of light capture when the wavelengths of the exciting light and the Stokes component value were near the photonic stop- band edges. Accordingly, for a diameter of globules D=260 nm, the spectral position of water infiltrated photonic crystal stop-band corresponds to the value of λ_{max} = 607.0 nm. The spectral width of the corresponding stop-band is about 70 nm. Taking into account that in this case the spectral position of the first Stokes component corresponds to 650 nm and exciting line – to 532 nm, we have the conditions of the first Stokes threshold lowering. Such theoretical prediction is in agreement with obtained experimental results. Similar effect was observed for nitrobenzene and ethanol filled opal matrices with corresponding globules diameters.

We have also shown SRS threshold lowering as a result of light capture in liquid filled photonic glasses, constructed from spherical globules with diameters of micro- or millimeter scales size. In this case SRS threshold lowering may result from resonance Mie [13] or whispering gallery modes.

Thus we have concluded that SRS threshold of liquids embedded into photonic structures may be essentially lowered with comparison to pure liquids.

Acknowledgments

This work was supported by RFBR (grants 12-02-00491, 13-02-00449, 13-02-90420, 14-02-00190, 14-02-90406, 14-02-90007).

References

[1] Herzberg G 1966 *Molecular Spectra and Molecular Structure* (Princeton: D Van Nostrand Comp.)
[2] Walrafen G E 1964 *J. Chem. Phys.* **40** 3249-56
[3] Eisenberg D and Kauzman W 1969 *The Structure and Properties of Water* (Oxford: Clarendon Press)
[4] Lawandy N M, Balachandran R M, Gomes A S L and Sauvain E 1994 *Nature* **368** 436 - 38
[5] Sha W L, Liu C- H and Alfano R R 1994 *Opt. Lett.* **19** 1922-24
[6] Noginov M A, Noginova N E, Caulfield H. J. et al. 1996 *J. Opt. Soc. Am. B.* **13** 2024-33
[7] Astratov V N, Bogomolov V N, Kaplyanskii A A, Prokofiev A V, Samoilovich L A, Samoilovich S M and Vlasov Yu A 1995 *Nuovo Cimento D* **17**, 1349 - 54
[8] Gorelik V S, Zlobina L I, Sverbil' P P, Fadyushin A B and Chervyakov A V 2005 *Journal of Russian Laser Research* **26** 211-27
[9] Gorelik V S 2007 *Quantum electronics* 2007 **37** 409-32
[10] Gorelik V S 2008 *Laser Physics* **18** 1479-1500
[11] Gorelik V S , Kudryavtseva A D and Tcherniega N V 2008 *Journal of Russian Laser Research* **29** 551-57
[12] Gorelik V S 2010 *The European Physical Journal Applied Physics* **49** 33007
[13] Mie G 1908 *Ann. Phys.* **25** 377-445

Numerical simulation of wall relaxation in Rb vapor optically pumped with ^{21}Ne

D A Dolgopyatov, E S Navasardyan and I A Arkharov

Bauman Moscow State Technical University, 2nd Baumanskaya St., 5, Moscow, 105005, Russia

E-mail: arkharov@bmstu.ru

Abstract. Spatial distribution of the polarization in Rb vapor optically pumped with ^{21}Ne has been modeled numerically with 40 mW laser beam power, 180°C cell temperature, 1 bar cell pressure, and 6 mm × 15 mm pancake-cell geometry. An inhomogeneity of polarization throughout the cell results from a wall relaxation of polarized media and a decay of the pumping beam. For small cells characteristic of experimental conditions, a contribution of wall relaxation to the total alkali relaxation rate can become significant, demanding more accurate evaluation than afforded by analytical model based on average parameters.

1. Introduction

Isotopes of noble gases with non-zero nuclear spin—^{3}He, ^{21}Ne, ^{83}Kr, ^{129}Xe, ^{131}Xe—can be utilized for nuclear magnetic resonance (NMR) imaging in medical or scientific applications. This work on ^{21}Ne has been expedited by a ^{3}He supply crisis, opening the way to alternative biomarkers, as well as by the progress in the cryogenic technology of isotope enrichment [1].

To increase signal-to-noise ratio, the nuclear spins of isotopes should be hyperpolarized well above thermal equilibrium under conditions of tomographic magnetic field strength. This is performed with spin-exchange optical pumping (SEOP) of noble gas isotopes in the presence of alkali vapor [2].

In the present paper, we numerically simulate the steady-state polarization of Rb vapor, optically pumped with ^{21}Ne buffer gas, accounting for its 2D spatial distribution caused by wall relaxation, and compare the results with an analytical solution based on a wall relaxation rate averaged throughout the cell. The media properties for the simulation were taken from [3]–[6] .

2. Theory

SEOP is a process of angular momentum transfer from circularly polarized light to alkali (Rb) electron polarization of Zeeman sub-levels and via hyperfine interactions to noble gas (^{21}Ne) nuclear polarization of Zeeman sub-levels. The alkali nuclei also become polarized. These traits can be modeled with the following equations [6]–[7].

Equilibrium alkali polarization equation:

$$P^a = \frac{\frac{1}{\epsilon^a+1} \cdot R_{op}}{\frac{1}{\epsilon^a+1} \cdot R_{op} + R^a_{rel}} \tag{1}$$

Content from this work may be used under the terms of the Creative Commons Attribution 3.0 licence. Any further distribution of this work must maintain attribution to the author(s) and the title of the work, journal citation and DOI.
Published under licence by IOP Publishing Ltd

Pumping beam decay equation:

$$\frac{dR_{op}}{dz} = -n^a \sigma(\omega)(1 - P^a) R_{op} \tag{2}$$

Equilibrium noble gas polarization equation:

$$P^b = P^a \frac{\epsilon^b}{I^b/S} \frac{R_{se}^{ba}}{R_{se}^{ba} + 1/T1^b} \tag{3}$$

Their parameters are further given by:

$$R_{rel} = \frac{1}{\epsilon^a + 1}(R_{sd}^{aa} + R_{sd}^{ab} + R_{se}^{ab} + R_{pr}) + R_{wall}^a \tag{4}$$

$$R_{sd}^{aa} = n^a \sigma_{sd}^{aa} v^{aa} \tag{5}$$

$$R_{sd}^{ab} = n^b \sigma_{sd}^{ab} v^{ab} \tag{6}$$

$$v^{ij} = \left(\frac{8kT}{\pi M}\right)^{1/2} \tag{7}$$

$$M = \left(\frac{1}{m^i} + \frac{1}{m^j}\right)^{-1} \tag{8}$$

$$R_{se}^{ab} = R_{se}^{ba} \frac{n^b}{n^a} \tag{9}$$

$$R_{se}^{ba} = \kappa_{se}^{ab} n^a \tag{10}$$

$$R_{wall}^a = D^a \left(\frac{\pi}{r}\right)^2 \tag{11}$$

$$\epsilon(I, \beta) = (2I + 1)\mathrm{cth}(\beta/2)\mathrm{cth}[\beta(I + 1/2)] - \mathrm{cth}^2(\beta/2) \tag{12}$$

$$\beta = \frac{1 + P}{1 - P} \tag{13}$$

Here, superscripts a and b refer to alkali and noble gas, respectively; i and j to species of corresponding type; P is polarization; n is atomic density, m^{-3}; R_{op} is the optical pumping rate, s^{-1}; R_{rel} is the relaxation rate, s^{-1}; R_{sd} is the spin destruction rate, s^{-1}; R_{se} is the spin exchange rate, s^{-1}; R_{pr} is the probe-beam pumping rate, s^{-1}; R_{wall} is the wall relaxation rate, s^{-1}; $T1$ is a longitudinal relaxation time constant, s; z is a coordinate along pumping beam axis, m; σ is the photon-absorption cross section, m^2; ω is the pumping beam frequency, Hz; σ_{sd} is the spin-destruction cross section, m^2; v is the relative velocity of colliding particles, m/s; k is the Boltzmann constant; T is cell temperature, K; M is the reduced mass of colliding particles, kg; m is the mass of a particle, kg; D is the diffusion coefficient, m^2/s; I is nuclear spin; S is electron spin; ϵ is a paramagnetic coefficient; β is spin temperature, r is the spherical cell radius, m; and κ_{se} is the spin-exchange-rate coefficient, m^3/s.

This model neglects the relaxation due to magnetic-field inhomogeneity of the cell and the presence of quenching nitrogen, N_2. The probe beam is not considered further in this paper, thus $R_{pr} = 0$.

The model above has the following convenient analytical solution, which permits the calculation of 1D polarization distributions of Rb and ^{21}Ne using equations (1) and (3):

$$R_{op}(z) = (\epsilon^a + 1) R_{rel}^a \cdot \mathrm{W}\left[\frac{1}{\epsilon^a + 1} \cdot R_{op-int} \frac{\exp\left[\frac{\frac{1}{\epsilon^a+1} \cdot R_{op-int}}{R_{rel}^a} - \sigma(\lambda_{D1}) n^a \cdot z\right]}{R_{rel}^a}\right] \tag{14}$$

$$R_{op-int} = \frac{\sigma(\lambda_{D1}) P_{Pump} \lambda_{D1}}{Ahc} \tag{15}$$

where W is a Lambert W-function; λ_{D1} is the Rb D1-transition wavelength, m; P_{Pump} is pumping-beam power, W; A is the cell cross-sectional area, m^2; h is Planck's constant; and c is the speed of light.

One can evaluate the normalized contribution of the wall relaxation rate R^a_{wall} for different temperatures, pressures, and cell radii. Figure 1 shows a sample of such calculations. The normalized contribution of wall relaxation is significant for low buffer-gas pressure and small cell size, which can be the case in laboratory experiments. For example, under the conditions simulated, the R^a_{wall} is as high as 34.3% the total alkali relaxation rate, assuming an equivalent cell radius

$$r = \frac{3V}{A} \tag{16}$$

where V is cell volume, m^3.

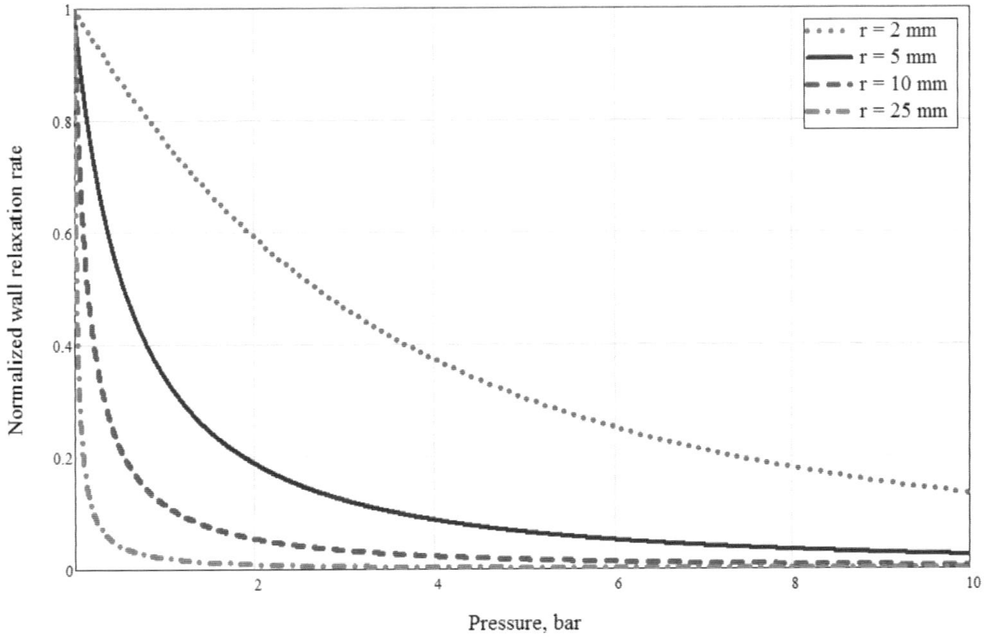

Figure 1. Normalized wall relaxation rate for spherical cells with radii of 2 to 25 mm.

The evaluation of R^a_{wall} by equation (14) provides a rough estimate of its effective average value, which is even less precise for non-spherical cells. To account for wall relaxation more accurately, equation (1) should be substituted by the alkali diffusion equation:

$$D^a \nabla^2 P^a + \frac{2Q^a}{n^a} = 0 \tag{17}$$

$$Q^a = n^a \left(\frac{1}{\epsilon^a + 1} R_{op}(1 - P^a) - R^a_{rel} P^a \right) \tag{18}$$

$$R^a_{rel} = \frac{1}{\epsilon^a + 1} (R^{aa}_{sd} + R^{ab}_{sd} + R^{ab}_{se} + R_{pr}) \tag{19}$$

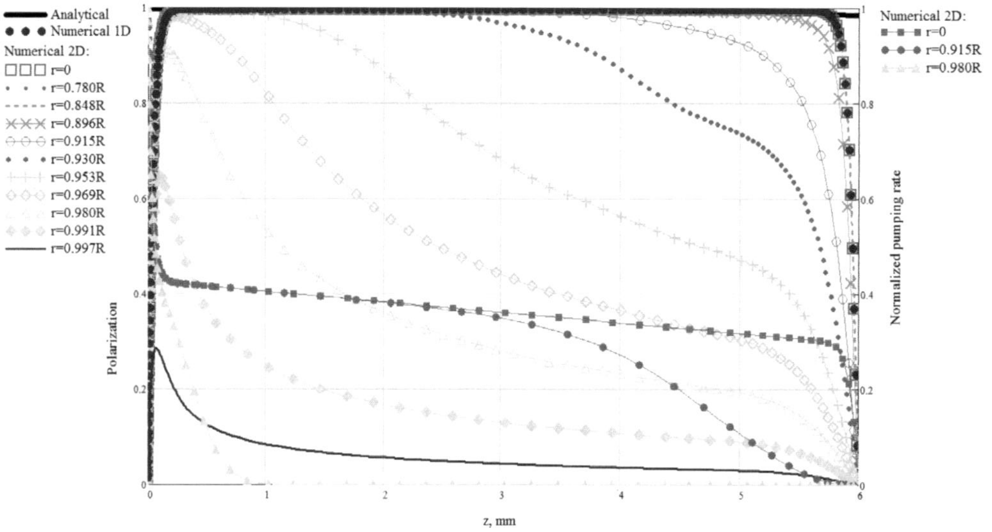

Figure 2. Results of numerical simulation: Rb polarization and normalized pumping rate vs. z-coordinate.

where Q is the intensity of polarized particle sources, $m^{-3}s^{-1}$. Equation (17) has been solved numerically with the following boundary conditions:

$$D^a \left. \frac{\partial P^a(r,z)}{\partial z} \right|_{z=0} = \alpha^a P^a(r,0) \tag{20}$$

$$D^a \left. \frac{\partial P^a(r,z)}{\partial z} \right|_{z=L} = -\alpha^a P^a(r,L) \tag{21}$$

$$\left. \frac{\partial P^a(r,z)}{\partial r} \right|_{r=0} = 0 \tag{22}$$

$$D^a \left. \frac{\partial P^a(r,z)}{\partial r} \right|_{r=R} = -\alpha^a P^a(R,z) \tag{23}$$

$$\alpha = \frac{1}{4} \left(\frac{8kT}{\pi m^a} \right)^{1/2} \tag{24}$$

where r is the cell radial coordinate, m; and L is the cell length, m.

Equations (20) and (21) represent the wall relaxation at cell edges, equation (22) – a condition of symmetry on the cell axis, and equation (23) – wall relaxation on the cell's outer surface. Equation (24) was derived from the average wall collision frequency per particle, assuming a Maxwell distribution. Equation (2) was used for pump beam intensity calculations.

3. Results

Figure 2 presents the results of numerical simulation of Rb vapor polarization under 40 mW laser beam power, 180°C cell temperature, 1 bar cell pressure, 6 mm × 15 mm pancake-cell geometry.

The average simulated polarization of Rb across the cell is 91.3%, versus a value of 99.4% predicted by the analytical formula. These correspond to cell-average ^{21}Ne polarizations of 7.5% and 8.2%, respectively. 1D numerical simulation, neglecting the relaxation on the cell's outer surface, gives 98% polarization of Rb and 8.1% polarization of ^{21}Ne.

Journal of Physics: Conference Series **584** (2015) 012027 doi:10.1088/1742-6596/584/1/012027

4. Conclusion and future work

The simulated Rb and ^{21}Ne polarizations are in fair agreement with those predicted analytically using cell-average values for the wall relaxation rate R^a_{wall}. The error reaches 8.9% and contains a significant contribution of the relaxation on the cell's outer surface. 1D numerical simulation gives 1.4% difference with analytical solution and includes the wall relaxation at cell edges only. Minor corrections are possible in terms of accounting for neglected effects. However, it is recommended that numerical simulations should be performed for laboratory-like conditions, especially in cases with small cells and low buffer-gas pressures. The contribution of R^a_{wall} vanishes with increasing cell size.

The numerical simulation does exhibit differing qualitative results compared with the analytical formula. The boundary layers feature a marked decrease of polarization with a corresponding pumping-light absorption. This is especially important for the front edge of the cell, where the pumping beam's power can be completely diminished under certain conditions. Within the cell core, however, the polarization is fairly uniform.

Future work will include extensive simulation of ^{21}Ne polarization with K, Rb, and Cs under different cell conditions and with varying noble-gas isotope enrichments and modeling of transient effects.

References

[1] Arkharov A M, Arkharov I A, Dolgopyatov D A, and Bondarenko V L 2013 ^{3}He supply crisis: reasons and challenges *Chemical and Petroleum Engineering* **49** 41–45

[2] Conradi M S, Saam B T, Yablonskiy D A and Woods J C 2006 Hyperpolarized ^{3}He and perfluorocarbon gas diffusion MRI of lungs *Progress in Nuclear Magnetic Resonance Spectroscopy* **48** 63–83

[3] Alcock C B, Itkin V P and Horrigan M K 1984 *Canadian Metallurgical Quarterly* **23** 309

[4] Allred J C, Lyman R N, Kornack T W and Romalis M V 2002 High-Sensitivity Atomic Magnetometer Unaffected by Spin-Exchange Relaxation *Phys. Rev. Lett.* **89** 130801

[5] Franz F A and Volk C 1976 Spin relaxation of rubidium atoms in sudden and quasimolecular collisions with light-noble-gas atoms *Phys. Rev.* A **14** 1711–28

[6] Ghosh R J and Romalis M V 2010 Measurement of spin-exchange and relaxation parameters for polarizing ^{21}Ne with K and Rb *Phys. Rev.* A **81** 043515

[7] Walker T G and Happer W 1997 Spin-exchange optical pumping of noble gas nuclei *Reviews of Modern Physics* **69** 629–642

Experimental study of influence of nonlinear effects on phase-sensitive optical time-domain reflectometer operating range

E T Nesterov[1], A A Zhirnov[1], K V Stepanov[1], A B Pnev[1], V E Karasik[1], Ya A Tezadov[2], E V Kondrashin[2] and A B Ushakov[2]
[1] Bauman Moscow State Technical University, 2-nd Baumanskaya st. 5, 105005 Moscow, Russian Federation
[2] Scientific and Technological Enterprise IRE-Polyus, pl. Vvedenskogo 1, str. 3, 141190 Fryazino, Moscow region, Russian Federation

E-mail: evgeny.t.nesterov@gmail.com

Abstract. This paper presents experimental results of a study the influence of nonlinear effects on the range of the phase-sensitive optical time-domain reflectometer. Results from measurements of the impact of the effects of stimulated Raman scattering, stimulated Brillouin scattering, self-phase modulation, and modulation instability are presented. It is shown that the power of the probe pulse for 200 ns does not exceed ≈300 mW.

1. Introduction

Interest in distributed fiber-optic sensors, including a phase-sensitive reflectometer such as the Rayleigh optical time-domain reflectometer (OTDR), has increased over the past two decades. One of examples of a phase-sensitive OTDR – is a vibration sensor. Distributed optical fiber vibration sensors have advantages such as an extensive operating range, a single-ended fiber connection to the interrogator, and the ability to make additional connections to a pre-existing fiber-optic cable. In some cases, the last advantage listed makes a solution cost-effective. Digital electronics development in recent years has enabled practical realization of fiber-optic distributed sensors based on scattering effects [1–12].

The main advantage of distributed fiber-optic sensors is a large operating range. One of the factors limiting the range of the phase-sensitive OTDR is related to the power pulses input into the fiber that are due to the influence of nonlinear effects in the fiber [9, 11, 15-22].

The range of the distributed vibration sensor based on the phase-sensitive Rayleigh OTDR is determined by providing the desired signal-to-noise ratio (SNR) of the system in a given area of the optical fiber. An increase in the SNR can be obtained by decreasing the system's noise power or by increasing the supplied optical probe pulse energy to the fiber. In practice, increasing the probe pulse energy supplied to the optical fiber impulse power (at a fixed duration) leads to a decrease in the power of the backscattered signal. This phenomenon can be explained by appearance of nonlinear effects in optical fiber [14].

In this paper, we present experimental results of a study on the influence of nonlinear effects on the range of a phase-sensitive OTDR. We also present the results of measuring the impact of the effects of stimulated Raman scattering (SRS), stimulated Brillouin scattering (SBS), self-phase modulation (SPM), and modulation instability (MI). The results also show (Figure 2) that the power of the probe pulse duration of 200 ns does not exceed 300 mW.

Content from this work may be used under the terms of the Creative Commons Attribution 3.0 licence. Any further distribution of this work must maintain attribution to the author(s) and the title of the work, journal citation and DOI.
Published under licence by IOP Publishing Ltd

2. Experiments

To determine the dependence of reflectogram quality on probe pulse power, we built the experimental setup shown in Figure 1.

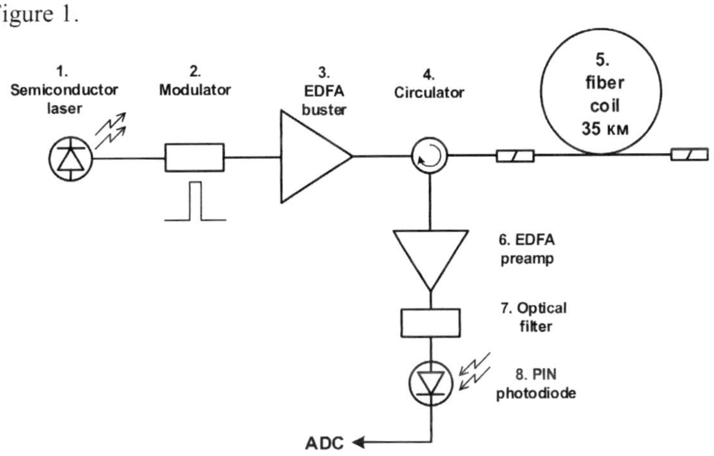

Figure 1. Experimental setup of COTDR: 1 is a light source, 2 is a modulator, 3 is an erbium-doped fiber amplifier (EDFA)-buster, 4 is an optical circulator, 5 is a fiber coil, 6 is an EDFA-preamplifier, 7 is an optical filter, 8 is a PIN photodiode.

Increasing the probe pulse power up to 250 mW (Figure 2 a, b) resulted in an increase of the reflectogram amplitude. Further increasing the power leads to a decrease in reflectogram amplitude, which starts from the end of the connected optical fiber sensor (Figure 2 c, d).

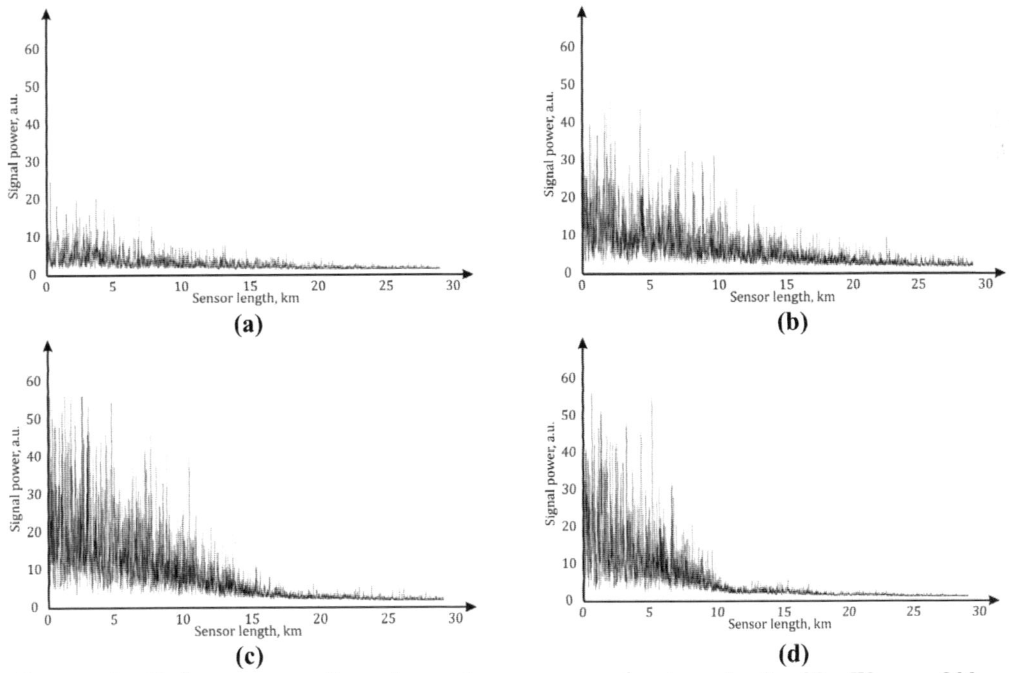

Figure 2. Reflectogram of probe pulse corresponds to: a) P = 80 mW, τ = 200 ns, b) P = 210 mW, τ = 200 ns, c) P = 280 mW, τ = 200 ns, d) P = 320 mW, τ = 200 ns.

Journal of Physics: Conference Series **584** (2015) 012028 doi:10.1088/1742-6596/584/1/012028

The decreasing amplitude can be explained by the appearance of nonlinear effects in the silica fiber. The following nonlinear effects are known to be present in the fiber [14]: SRS, SBS, SPM, and four-wave mixing (FWM), which in the case of the signal and the spontaneous emission radiation interaction is called MI. In the next sections, we assess the impact of non-linear effects in the Rayleigh COTDR.

3. Stimulated Raman scattering

SRS is manifested when incident radiation causes radiation generation with a shifted frequency, called Stokes radiation. The magnitude of the frequency shift is determined by the vibrational modes of the medium [14-16].

The probe optical pulse serves as a pump for generating radiation with a shifted frequency in the case of the pulse reflectometer. The weighted average maximum frequency shift is 13.2 THz with respect to the pump radiation frequency. The SRS threshold is defined as the input pump power that generates a Stokes radiation power equal to the output pump power.

Under the assumption of a Lorentzian shape of the gain spectrum, the critical pumping power can be estimated approximately by the formula [15]:

$$P_{SRS} \approx \frac{16A_{eff}}{g_{SRS}L_{eff}} \quad (1)$$

where P_{SRS} is the threshold power, $A_{eff} = 75\mu m^2$ is the effective mode area, $g_{SRS} = 6{,}5\text{x}10^{-14}\, m/W$ is the gain factor [5], and $L_{eff} = (1 - e^{-\alpha L})/\alpha$ is the effective fiber length; for typical losses in fiber of 0.2 dB/km, $\alpha \approx 4.8\text{x}10^{-5}\, m^{-1}$ and $L_{eff} = 18km$. Calculation according to Formula 1 gives the result:

$$P_{SRS} \approx 700mW \quad (2)$$

Our experimental setup for the SRS measurement is shown in Figure 3.

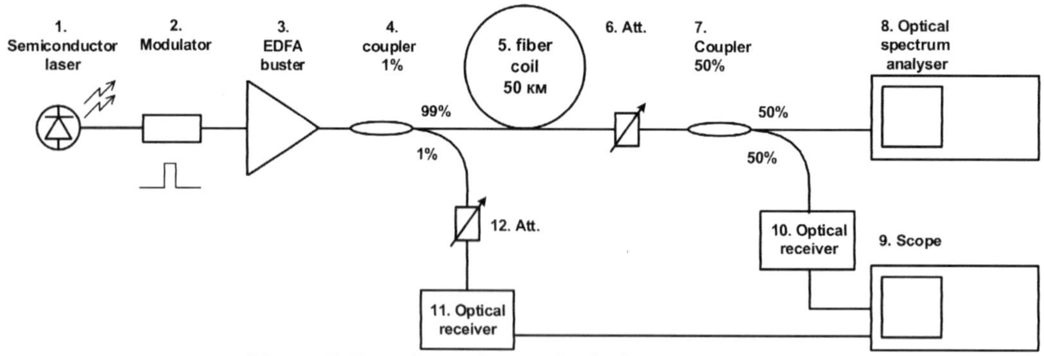

Figure 3. Experimental setup for SRS measurement.

Some of the obtained results of the measurements are shown below in Figures 4(a-d).

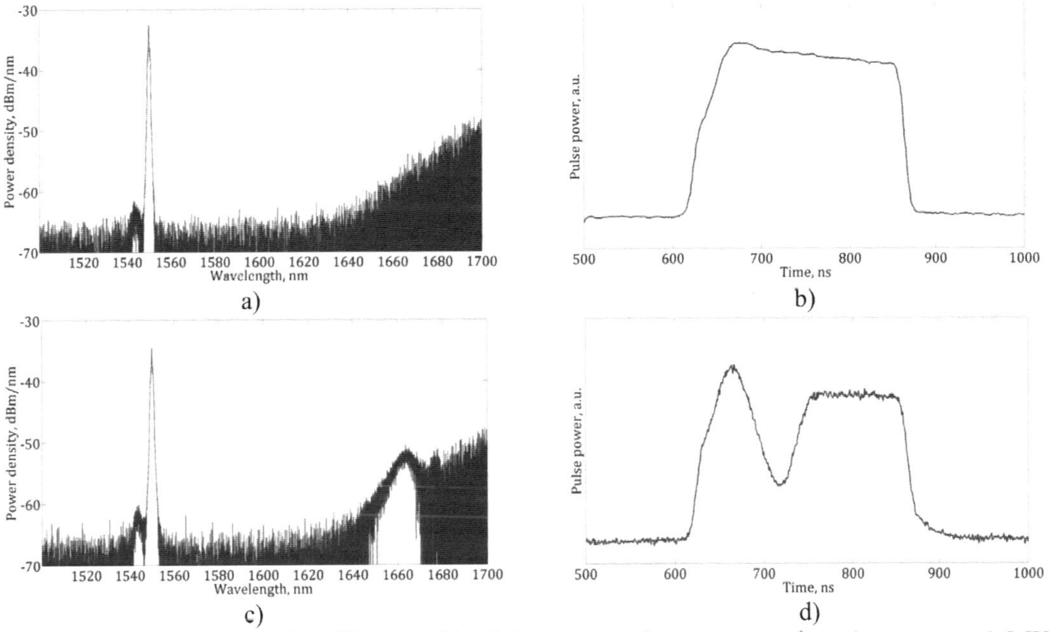

Figure 4. Spectrum (a) and oscillogram (b) of the output pulses corresponds to input power 1.5 W, pulse duration 200 ns. Spectrum (c) and oscillogram (d) of the output pulses corresponds to input power 2 W, pulse duration 200 ns.

As can be seen from Figures 4a-4d SRS begins to develop when the probe pulse power exceeds 1500 mW. That value is two times larger than the theoretical threshold previously calculated (2). The discrepancy between the theoretical calculation and experimental data can be explained by the fact that expression (1) is obtained for the case of continuous pumping. In the case of pulse pumping, the Stokes component is caused by the finite duration pulse. The difference between the Stokes emission wavelength and pump wavelength is about 100 nm (figure 4c). Due to the dispersion component in the fiber $D = 17 \text{ps} \times \text{nm}^{-1} \times \text{km}^{-1}$, after passing 50 km of fiber the Stokes wave and the probe pulse after passing undergo an 85-ns time delay (figure 4d).

The increasing of the SRS threshold to a value greater than 1.5 W is explained by finite pulse duration and time delay between probe pulse and generated Stokes wave. The obtained results indicate that SRS is not the main factor of the decrease in the reflectogram amplitude.

4. Stimulated Brillouin scattering

SBS is manifested in the form of Stokes wave propagation in the opposite direction with respect to the pump wave and containing a significant portion of the initial energy. The SBS process can be described classically as a parametric interaction between the pump, Stokes waves, and acoustic waves. Due to electrostriction, a pump wave generates an acoustic wave resulting in a periodic modulation of the refractive index. Induced refractive-index grating reflects the pump radiation [14-16].

The width of the SBS gain spectrum is about 10 MHz. The frequency shift of the reflected wave caused by the Doppler effect is 10 GHz. Only backscatter propagation of SBS is possible in a single-mode fiber. The SBS threshold is determined by the formula [17]:

$$P_{SBS} \approx \frac{21A_{eff}}{g_{SBS}}\frac{n}{Tc}, \tag{3}$$

where P_{SBS} is the threshold power, $A_{eff} = 75\mu m^2$ is the effective mode area, $g_{SBS} = 4{,}5x10^{-11}\ m/W$ is the gain [18], T = 200 ns is the pulse duration, $c = 3 \times 10^8\ m/s$ is the speed of light in vacuum, n = 1.5 is the refractive index of quartz. Using the expression given in (3), we find $P_{SBS} \approx 850$ mW.

The experimental setup for a study of the SBS influence is shown in figure 5.

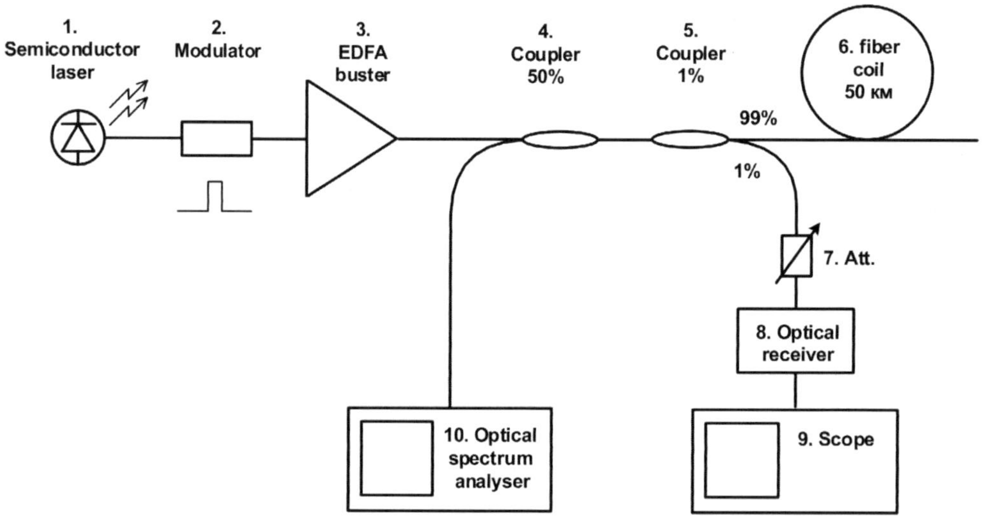

Figure 5. Experimental setup for SBS measurement.

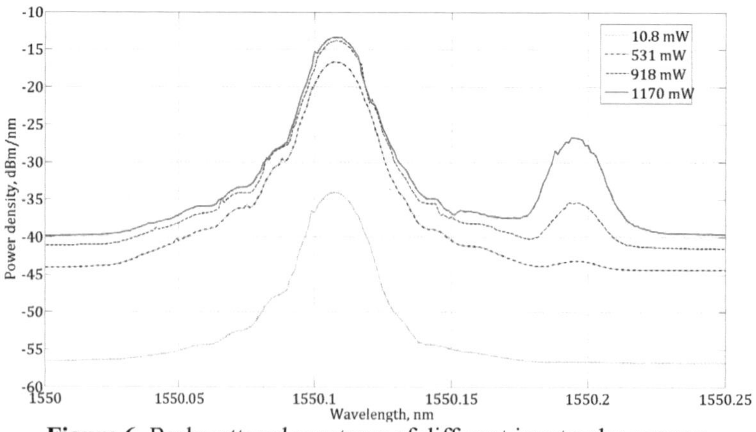

Figure 6. Backscattered spectrum of different input pulse powers.

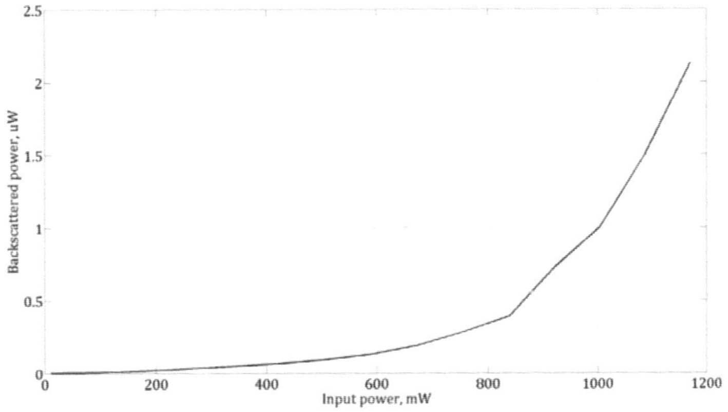

Figure 7. Backscattered power.

On Figure 6 and Figure 7, we can see that SBS begins to develop after the probe pulse power reaches a value of about 850 mW. Hence, SBS is not a main effect leading to a decrease in the reflectogram amplitude as the power of the probe pulse increases.

5. Self-phase modulation

SPM appears to be due to the dependence of the refractive index of quartz from the radiation energy injected into the fiber [14]. SPM causes a frequency shift in the scanning pulse passed through the fiber and can be calculated by equation (4) [19]:

$$\delta v = \frac{n_2 L_{eff}}{A_{eff}\lambda}\frac{\delta P}{\delta t}, \quad (4)$$

where n_2 is the nonlinear refractive index, L_{eff} is the effective fiber length, $A_{eff} = \pi a_m^2$ is the effective area of the fiber, a_m is the mode radius, λ is the emission wavelength, and P is the power of the optical pulse.

Backscattered Rayleigh radiation contains a frequency shift, which develops in a pulse as it passes through the optical fiber. Backscattered Rayleigh radiation (given to the band receiver) is proportional to the integral of the instantaneous power spectrum over the frequency [20], given by equation (5):

$$F(\delta v) = \int_{-1/T}^{1/T}\left[\frac{\sin(\pi(v-\delta v)T)}{\pi(v-\delta v)T}\right]^2 dv, \quad (5)$$

where v is the frequency, T is the pulse duration, $\Delta F = 1/T$ is the receiver bandwidth, and δv is the frequency shift caused by SPM. The decrease in sensitivity due to the frequency shift is proportional to:

$$d(\delta v) = F(\delta v)/F(0) \quad (6)$$

Figure 8 presents sensitivity reduction as a function of the ratio of the frequency shift to the electric band receiver tract. The electric band of the receiver tract is determined on the basis of the pulse duration. Therefore, for a pulse duration of $T = 200$ ns , the receiver bandwidth is $\Delta F = 5$ MHz

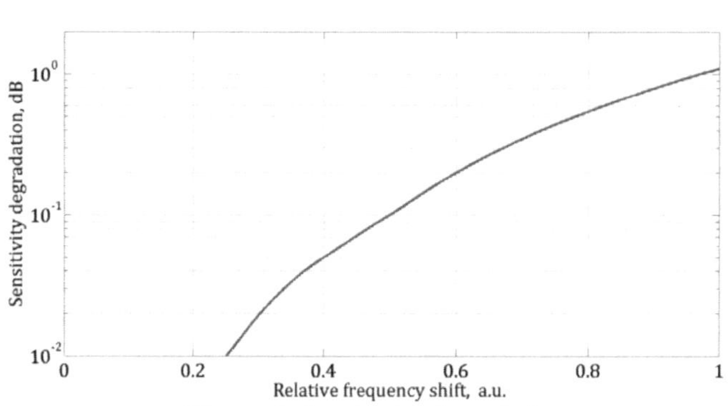

Figure 8. Sensitivity degradation

Our experimental setup was made for the quantitative measurement of the frequency shift $\partial\nu$. The scheme of the setup is shown in Figure 9.

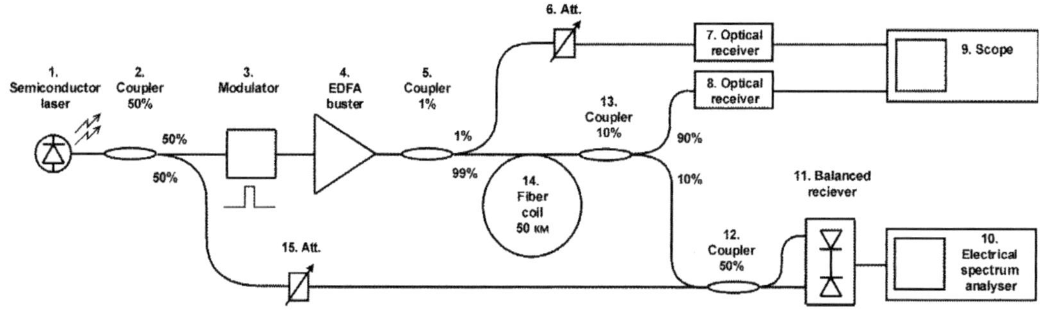

Figure 9. Experimental setup for study of the SPM influence.

The measurement results are shown in figures 10a and 10b below.

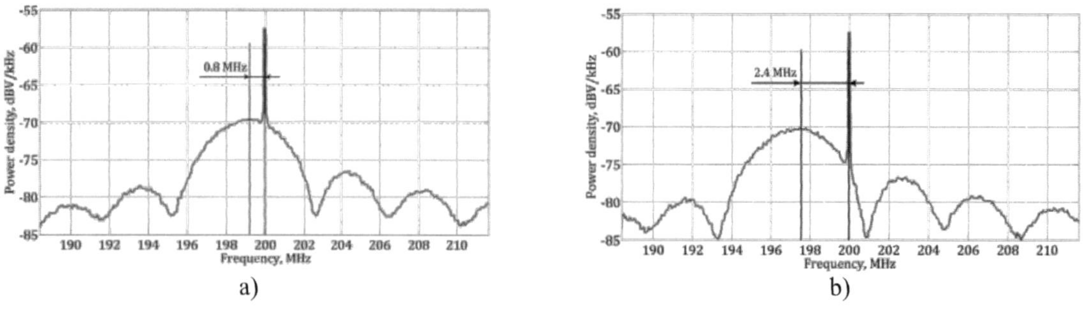

Figure 10. Signal spectrum corresponds to input pulse power: a) 300 mW and b) 500 mW.

As it shown in these figures, if the probe pulse power is 500 mW, then the frequency shift is 2.4 MHz. The sensitivity degradation is 0.1 dB, with $\partial\upsilon/\Delta F \cong 0.5$. Thus, SPM is not the main nonlinear effect limiting the power of the probe pulse injected into the fiber.

6. Modulation instability

MI is a degenerate case of FWM, representing a parametric amplification of noise [14, 15, 21, 22]. Pulses with power levels of several hundred milliwatts must be input into the optical fiber to ensure a sufficient SNR for backscatter signal processing. Such power levels can be achieved by using an EDFA. Optical amplifiers generate their own spontaneous emission noise. The interaction between signal and spontaneous radiation in the same frequency band, accompanied by the generation of new spectral components, is a degenerate case of FWM [14, 15].

Generation of new spectral components occurs at frequencies symmetrically shifted towards lower and higher frequencies. If the radiation frequency is indicated as ν_0, the new spectral components are generated at frequencies $\nu_0 + \nu_S$ and $\nu_0 - \nu_S$. The spectral component of the $\nu_0 + \nu_S$ and $\nu_0 - \nu_S$ frequency is called the anti-Stokes and Stokes component, respectively. The value of ν_S is determined by equation (7) [14, 21]:

$$\nu_S = \frac{1}{2\pi}\sqrt{\frac{2\gamma P_i}{|\beta_2|}}, \tag{7}$$

where P_i is the optical power, $\gamma = 2\pi n_2/(\lambda A_{eff})$ is the nonlinearity parameter, $n_2 = 3.2 \times 10^{-20} m^2/W$ is the nonlinear refractive index, $\lambda = 1.5\mu m$ is the wavelength, $A_{eff} = 75\mu m^2$ is the effective mode area, and β_2 is the coefficient of the group velocity dispersion calculated using Equation (8).

$$\beta_2 = -\frac{\lambda^2}{2\pi c} D, \tag{8}$$

Also, $D = 17ps \times nm^{-1} \times km^{-1}$ is the dispersion coefficient and c is the speed of light, as before. Taking into account the above values, we can get

$$\nu_S = 0.12nm. \tag{9}$$

The experimental setup for modulation instability study is shown in Figure 11.

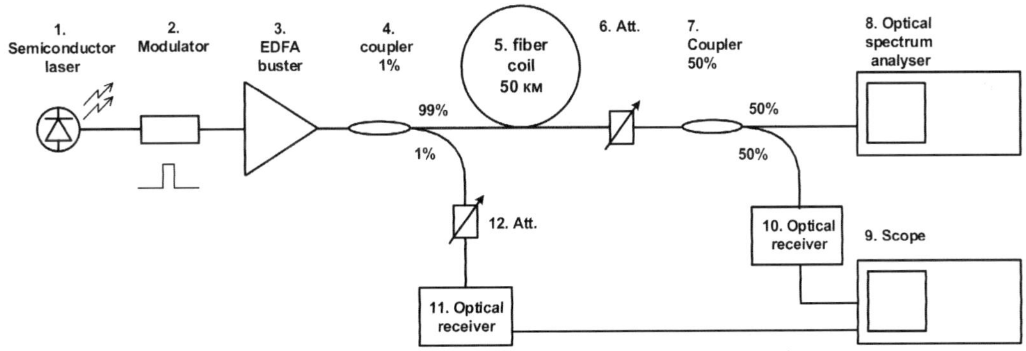

Figure 11. Experimental setup for calculation of backscattered power.

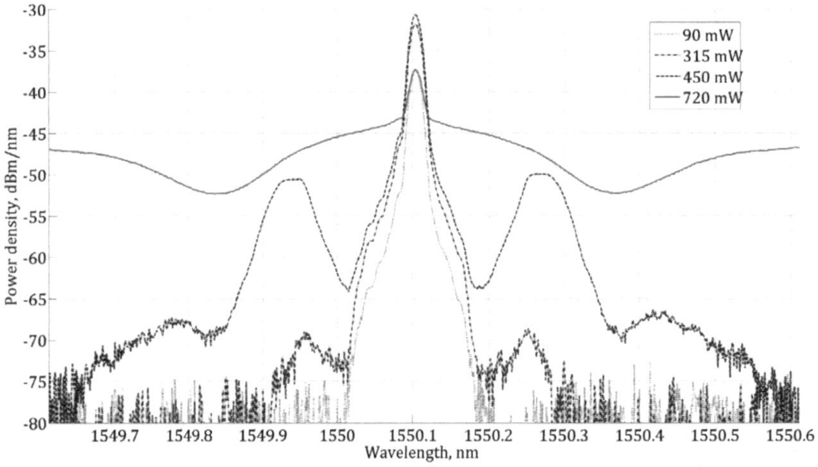

Figure 12. Output spectrum of different input pulse power.

The spectrum of a pulse passed through the 50 km fiber coil is shown in Figure 12. Results obtained from this data lead to the conclusion that Stokes lines appear when pulse power reaches the value of 315 mW. When pulse power reaches 450 mW, the power of the Stokes lines becomes commensurate with the power of radiation with the original input wavelength.

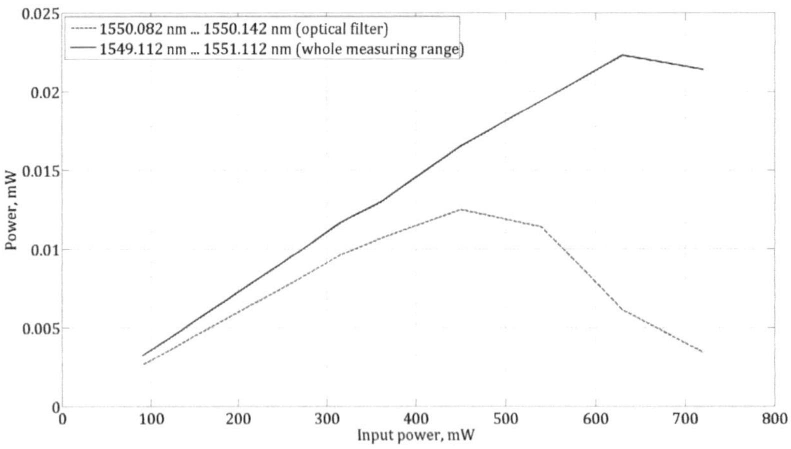

Figure 13. Output power in various ranges.

7. Discussion

An important feature of COTDR is the optical filter installed after pre-EDFA (see Figure 1). It has a bandwidth $\Delta\nu_F \approx 0.07$ nm and reduces the spontaneous emission noise of the fiber amplifier. Stokes lines wavelengths lay outside the range of this optical filter. That is the reason for the decrease in the optical power (Figure 13). As a result, the reflectogram amplitude decreases in the far fiber section (> 10 km) when the probe pulse power exceeds 300 mW.

8. Conclusions

Experimental results of the influence of four non-linear effects – stimulated Raman scattering (SRS), stimulated Brillouin scattering (SBS), self-phase modulation (SPM), and four-wave mixing (FWM)/modulation instability (MI) – on device operation with a scanning pulse duration of 200 ns are shown. We found that the main nonlinear effect is the MI, whose development is observed when the pulse power reaches 300 mW.

Acknowledgments
This work is supported by the Russian Ministry of Industry and Trade, State Contract № 13411.1007499.09.033. The authors are thankful to Treschikov V. N., Director of the 'T8'-company, Prof. Listvin V. N., and Prof. Naniy O. E. for useful discussions.

References

[1] Taylor H F and Lee C E 1993 *US 5194847 A*
[2] Choi K N, Juares J C and Taylor H F 2003 *Proc. SPIE* **5090** 134-41
[3] Juarez J C et al. 2005 *J. Lightwave Technol.* **23** 2081-87
[4] Juares J C and Taylor H F 2007 *Appl. Opt.* **46** 1968-70
[5] Juskaitis R, Mamedov A M, Potapov V T and Shatalin S V 1992 *Opt. Lett.* **17**, 1623
[6] Healey P and Malyon D J 1982 *Electron. Lett.* **18** 862-63
[7] Tai K, Hasegawa A and Tomita A 1986 *Phys. Rev. Lett.* **56** 135-38
[8] Martins H F, Martin-Lopez S and Corredera P 2013 *Opt. Lett.* **38** 872-74
[9] Lu Y L, Zhu T, Chen L A and Bao X Y 2010 *J. Lightwave Technol.* **28** 3243
[10] Izumita H, Koyamada Y, Furukawa S and Sankawa I, 1994 *J. Lightwave Technol.* **12** 1230
[11] Simaeys G V, Emplit P and Haelterman M 2001 *Phys. Rev. Lett.* **87** *033902*
[12] Simaeys G V, Emplit P and Haelterman M 2002 *J. Opt. Soc. Am. B* **19** 477
[13] Alasia D, Herraez M G, Abrardi L, Lopez S M and Thevenaz L 2005 *Proc. Soc. Photo-Opt. Instrum. Eng.* **5855** 587
[14] Agrawal G P 2013 *Nonlinear fiber optics* (New York: Academic Press)
[15] Izumita H, Koyamada Y, Furukawa S I and Sankawa I 1994 *J. Lightwave Technol.* **12** 1230-38
[16] Smith R G 1972 *Appl. Opt.* **11** 2489–94
[17] Koyamada Y, Nakamoto H and Ohta N 1992 *J. Opt. Commun.* **13** 127-33
[18] Stolen R and Rogers H 1980 *Proc. IEEE* **68** 1232-36
[19] Stolen R and Lin C 1978 *Phys. Rev.* **17** 1448
[20] Healey P 1984 *Electronics letters – IET* **20** 30-32
[21] Stolen R and Bjorkholm J 1982 *Quantum Electronics IEEE J.* **18** 1062-72
[22] Babin S A et al. 2010 *Laser Phys.* **20** 334-40

AUTHOR INDEX